Praise for *SuperSight*

"We're on the cusp of a step forward in our evolution as a species, and in *SuperSight,* David Rose provides a time-machine trip into the near future, where knowing more about anything is as simple as glancing at it."

—MIT Media Lab professor and head of the Fluid Interfaces group

"*SuperSight* gives us super insight into the new world of spatial, wearable, ubiquitous computing and its economic and social implications. And you've got to love an AR-enabled book. Like the future of humans, it's augmented!"

—Charlie Fink, technology columnist for *Forbes*

"Extended reality is poised to reinvent our human experience with technology, and *SuperSight* is an enlightening guide to the changes ahead."

—Paul Daugherty, CTO at Accenture

"David Rose has a wonderful way of making emerging technologies relatable and understandable. He gives insight into the layered future world that we are clearly headed towards. *Supersight* gives us a chance to proactively anticipate and prepare for the exciting new world of AI."

—Jordan Goldstein, Global Director of Design at Gensler

"We've been seeing augmented reality pop up in sci-fi movies for decades, from *Robocop* to *Minority Report* to *Iron Man.* But in this book, David Rose shares his clear and lively vision for how AR is actually going to change cities, business, and our brains. *SuperSight* is a must-read if you want to navigate augmented reality's transition from science fiction to a widely used technology."

—Scott Kirsner, CEO of Innovation Leader and technology columnist for *The Boston Globe*

"*SuperSight* is an elegant, informative, and thoughtful exploration of a future where our digital and spatial worlds merge seamlessly. We can never be certain about the future, but with foresight comes better preparedness. A delightful read."

—Jit Kee Chin, Chief Data and Innovation Officer at Suffolk

"A truly forward-looking book describing a new world in which the physical converges seamlessly with the ever-present digital—via personal, daily-wear smart glasses. Such AR glasses will require very innovative technologies, new modes of applications, and new human cognitive interaction. The opportunities and challenges here are both enormous. David Rose has leaped forward to provide us a good peek into such a future."

—Dr. John C. C. Fan, CEO of Kopin

"Drawing on his rich experience as a researcher and entrepreneur in spatial computing, David Rose paints a very clear and compelling future in everyday language about how AR and smart glasses will affect the lives of everyday people—in the very near future.""

—Ori Inbar, founder of Augmented World Expo

"David Rose sees the world in a uniquely informative way and communicates that vision crisply and clearly. His career trajectory shows us true leadership in the disciplines of computer science, design and entrepreneurship. All of that knowledge and ability culminates in this book. This is a timely release, as the discipline of computer vision is hitting its stride in the healthcare market but the potential for this technology is very much still unmet. Thus, for anyone who wants a lesson in this important topic or to gain insights from a truly exceptional mind and writer, this is must-read."

—Joseph Charles Kvedar, MD, professor of dermatology at Harvard Medical School and chair of the board, American Telemedicine Association

"We are at a Goldilocks moment in our collective computational history, where a personal computing experience transitions into a shared, spatial, and ubiquitous one. *SuperSight* is a timely and comprehensive journey into this next chapter of computing."

—Valentin Heun, PhD, VP of research at PTC

"In the ambitious forward-looking spirit of Kevin Kelley's *The Inevitable,* David Rose shows us where the arrows of technology are pointing. More importantly, he shows how these capabilities come together by telling stories of the personal, relatable experiences in our future."

—Craig Sampson, professor at Segal Design Institute at Northwestern University, and founder of IDEO Chicago

"Without a doubt, this book illuminates the future of computing. Computers have shrunk from room-sized mainframes to handheld devices with even more computational power. The next stage in this evolution is augmented smart glasses that fuse the digital world to the real one. Read *SuperSight* to understand what this shift will mean for the future of work and collaboration."

—Paul Travers, president and CEO of Vuzix Corporation

"Fascinating and timely. David Rose's first book, *Enchanted Objects,* showed us how smart everyday objects can change our lives. In *SuperSight,* he shows how augmented reality will radically alter the way we interact—not only with each other, but with the entire world around us."

—Hari Nair, consumer packaged goods innovation leader

"*SuperSight* gives us a critical, thought-provoking, and palpable view of the future of our reality. Rose blurs and refocuses the boundary between what is real and what is augmented, providing a guide to living in the inevitable simulacrum. This is essential reading for the next generation of changemakers."

—Hani Asfour, dean at Dubai Institute of Design and Innovation

"When you think of AR as your own personal genie or athletics coach perched tightly on the edge of your eyelashes guiding you exuberantly into the future as David Rose does in *Super-Sight,* the world looks much more exciting."

—Gilad Rosenzweig, head of MIT DesignX

"David Rose has a huge gift for sketching futuristic possibilities that will be upon us before we know it. The opening vision of *SuperSight* is breathtaking! And quite scary. Designers will be inspired by the new possibilities. To capture those possibilities, business leaders and policy planners should read this book to start figuring out how to guide SuperSight in a societally thoughtful way."

—Frank Gillett, veteran analyst at Forrester

"In his insightful new book *SuperSight,* David Rose gives us a visionary's look into the compelling possibilities (and potential pitfalls) of augmenting human capabilities by further superimposing our physical and digital worlds. This next wave of technology has the potential to transform how we interact with each other and with the environments around us, and to create exciting new opportunities for design."

—Ari Adler, Executive Director at IDEO

"*SuperSight* is a monumental invention, both promising and perilous for society."

—Joyce Sidopoulos, cofounder and vice president of Programs & Community at MassRobotics

"The world is about to change—again—and the impact will be larger than even the smartphone revolution. David Rose describes this next great wave of highly personal computing from his perspective as a front-line innovator and entrepreneur. *SuperSight* is an excellent guide to the many opportunities and hazards that await us."

— Jason McDowall, Creator and Host of the *AR Show* podcast and VP at Ostendo Technologies

SUPER SIGHT

Also by David Rose

Enchanted Objects

SUPER SIGHT

What Augmented Reality Means for Our Lives, Our Work, and the Way We Imagine the Future

DAVID ROSE

BenBella Books, Inc.
Dallas, TX

dlr@mit.edu
SuperSight.world

BenBella Books, Inc.
10440 N. Central Expressway
Suite 800
Dallas, TX 75231
benbellabooks.com
Send feedback to feedback@benbellabooks.com

BenBella is a federally registered trademark.

Printed in the United States of America
10 9 8 7 6 5 4 3 2 1

Library of Congress Control Number: 2021020936
ISBN 9781950665808
eISBN 9781637740125

Editing by Leah Wilson
Copyediting by James Fraleigh
Proofreading by Michael Fedison and Sarah Vostok
Indexing by WordCo Indexing Services
Text design and composition by Aaron Edmiston
Cover design by Chris McRobbie
Printed by Lake Book Manufacturing

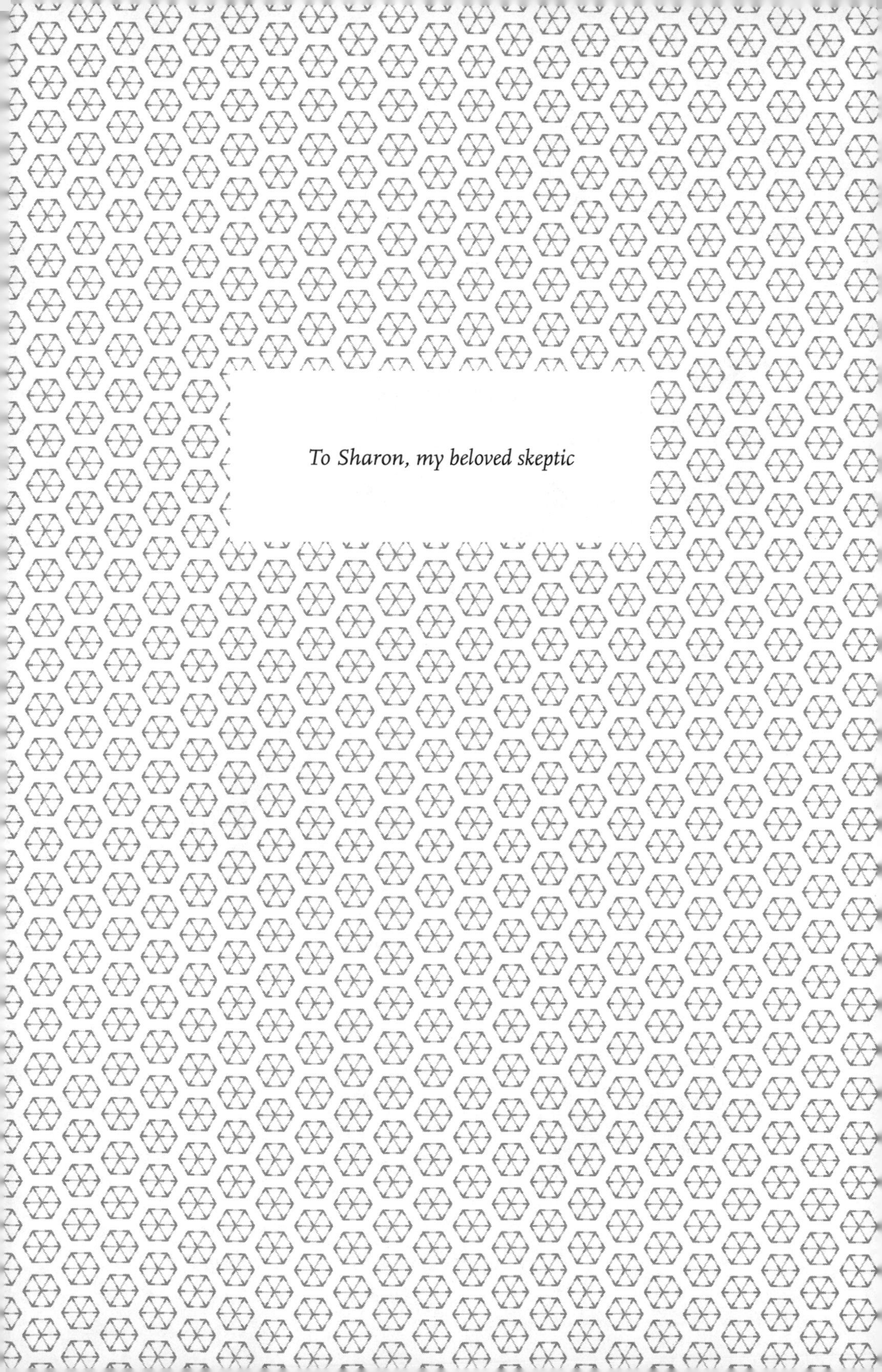

To Sharon, my beloved skeptic

CONTENTS

Part 2

Organizational Scale

Part 3

Society Scale

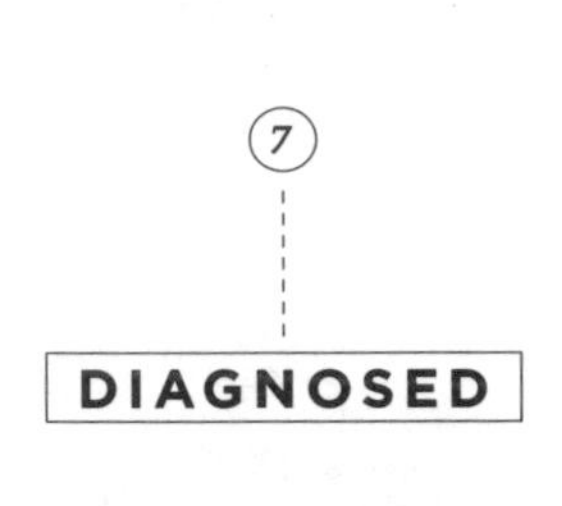

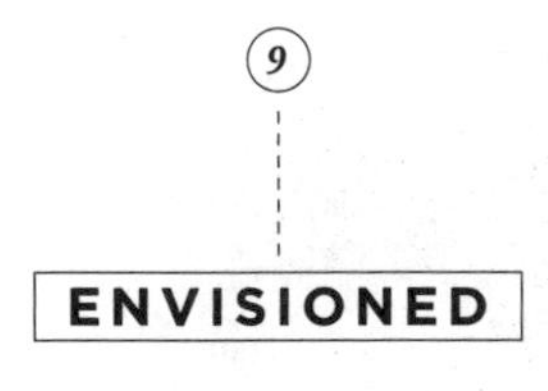

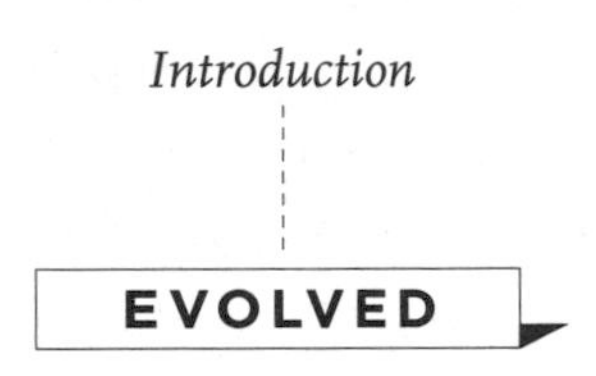

Introduction

EVOLVED

Imagine this: you are strolling down Lafayette Street in New York on a gloomy December morning, wearing your new augmented glasses. Thanks to the tiny data projector and optical combiner in these stylish spectacles, virtual and real are nearly indistinguishable. A holographic digital layer blends and "sticks" to the world as you move your head. And this new way of seeing is tailored just for you. The person walking next to you has a different curated projection.

The first thing you notice is how much richer and saturated with information the world has become. As you look up, the skyline includes translucent future buildings, some still in a sketch style to invite your feedback for the next zoning meeting, others rendered with detailed materials and flagged with their estimated completion date. Better not stare up at those high-rise residences too long or you'll start getting ads for their units, with their panoramic views, projected into your current apartment's windows.

So you look down. A widened sidewalk and cycling lane are superimposed on the street to show a redesign coming in a few months, with a graph projecting a drop in bike accidents. Following your system preferences, highly rated family-owned restaurants float huge, recommended dishes across your path. Here comes tortellini from the right; now a sushi boat from your left; and dead ahead, a steaming bowl of projected ramen—that's worth a quick stop.

The second, more problematic thing you notice is that these augmentations aren't neutral. They appear optimized to stimulate your particular brain and are biased toward positivity (or was that a configuration preference, too?). Your AI "reality editor" infers that you don't want to see the trash cans it detects lining the street this morning, so it replaces them with virtual bushes and trees transplanted from your childhood front yard. New York City never looked so good! (If only there were a search-and-replace for smells . . .) And because the glasses sense your mood through EEG (brainwave profiles) and EDA (electro-dermal activity) on your temples, and track your pupillary saccades (how your eyes subconsciously dart around to absorb a scene), what you see responds dynamically. Reacting to your sullen affect this morning, your glasses bend your reality, hoping to buoy your spirits: they brighten the overcast sky, add rays of summer sun, fade up some music with a beat that matches your footfalls, and include an affirmation in the now-blue sky, where clouds form your mantra: "You got this."

Third, you realize this new way of seeing is even more intensely social than Facebook. Instead of street advertising, billboards feature videos of friends and favorite celebrities speaking to you, translated into Spanish to help you polish up your vocabulary for an upcoming holiday. The sidewalk shimmers with embedded Hollywood Walk of Fame–style stars with names and faces of relatives and your best professors and mentors. What a nice way to stroll down memory lane.

Because you're always looking for a new jacket, your glasses highlight on-trend items others are wearing (as long as the manufacturers don't use child labor), and you bookmark them with a quick "eyebrow raise" gesture. Thanks to the TindAR smart glasses app, you automatically "edit out" any recent dates you happen to pass that didn't go so well—like that one who ghosted you last week (ouch!). Like in the *Black Mirror* episode, they appear, suitably, as a ghostlike figure. You said you never wanted to see them again; these glasses make it so.

Later, you go back out, this time for a run. Navigation is easy since you follow a yellow brick road painted for your eyes only. Yesterday's 10K had you running alongside the Scotsman from *Chariots of Fire*, then Rocky, then Usain Bolt for the final sprint—entertaining, but not as motivational as you needed. Today your glasses superimpose a pack of zombies chasing after you. They won't catch you if you exceed yesterday's pace, which you do! These glasses bend and edit reality, but they also project the future. As you run past a shop window, the reflected image of you is older: your hair is graying, but you wear a 2032 Olympics T-shirt, and you look lean, fast, and healthy.

I call this new visual reality *SuperSight.*

The human eye is an extraordinary organ. Packed with more than 120 million photoreceptor cells, it can discern 10 million different colors and has the body's fastest muscles—the average blink lasts just 100 milliseconds. It is also the most complicated organ after the brain, containing more than 2 million moving parts. For all of its wonders, however, the human eye hasn't significantly evolved in millennia. Although we've invented glasses to correct our vision, and microscopes and telescopes for specialized tasks, our ancestors perceived the world much as we do. But thanks to a set of exponentially advancing technologies over the next decade, that's about to change radically. We're going to experience an epic evolution in human sight, and I hope to conjure a vision for you of what this will feel like.

You're likely already familiar with *virtual reality*, or VR—immersive headsets like Oculus Quest or HTC's VIVE that transport their wearers to fantasy game worlds. But this book isn't about virtual reality. Those wearable screens *remove* us from reality; they're opaque and only enable experiences divorced from the world around you. SuperSight builds a new dimension on top of the existing world by spatially situating information on the ground-plane of reality. As Jim Heppelmann, the CEO of PTC, a software innovator in spatial computing, told me, "It's far more useful to decorate the existing world around us with information." This decoration, superimposed and pinned to the world, is often called *augmented reality*, or AR.

The pioneering research for AR was funded for military applications. At the US Air Force Research Lab in the early 1990s, engineers wanted to make it easy for a remote operator to control robot hands. They used an optical combiner to overlay the image of the robot's arms with the user's actual arms, and mixed computer-generated graphics with a view of the room to simulate physical barriers. With this mashup of digital information and the actual world, augmented reality was born. Later, aerospace engineers researching ergonomics found another reason to love AR: it allowed them to move the dashboard information of airplanes onto a window display, closer to the pilot's line of sight, thus reducing their cognitive load. These heads-up displays (HUDs) were followed by the development of helmets for air force pilots that use HUDs for targeting and landing guidance.

In 2016, Microsoft launched HoloLens, the first widely available AR headset. It used tracking technology originally developed for the Kinect gaming 3D sensor, which we had been using in academic interface research for some time, for depth sensing and body tracking. Around the same time, people in the industry started using the term *mixed reality* (MR) for these augmentations, because they were getting richer and more responsive. Two technologies drove this shift: (1) range-finding depth cameras that could read the dimensions of the world in front of the wearer; and (2) improvements in real-time graphics rendering that could position dynamic, interactive digital overlays in response to the environment. This meant virtual objects could be placed in specific physical places—sitting on a table, half-hiding behind a doorway, flying through a window—rather than just hovering in midair, as they did with the first-generation Google Glass.

New industries often have drifting nomenclature problems. For a brief period in 2018, we tried folding all these immersive technologies (VR, AR, and MR) under the single moniker XR. That was confusing, even for the small set of developers, investors, and journalists in this budding industry. Now we are starting to use a more descriptive term, *spatial computing*, to describe this new paradigm. The term *ambient computing* is also used. Both describe a future of immersive computing that fuses the physical world with digital augmentation. This is a big departure from the "anywhere, anytime" promise of the smartphone. Spatial computing places a particular emphasis on place and context. It organizes information—infinitely searchable, and therefore largely inaccessible—around something we already understand and come in contact with: the physicality of the world.

Snap and Instagram fans have already experienced one example of spatial computing's ability to anchor digital content on the physical world. You're using computer vision and a scaled digital projection every time you superimpose a cute koala nose, bunny ears, or barfing rainbows over your face, or animate a 3D Apple Animoji unicorn puppet by sticking out your tongue. Brands are also jumping in to adorn your face with real products you can purchase like Warby Parker glasses and Sephora makeup. Soon, salons will preview hairstyles as you walk past, and Swarovski will project earrings and a necklace that match your outfit. SmileDirectClub already gives you an animated preview of your "smile journey" after they've scanned the current arrangement of your teeth. Likewise, social media ads for "quick and painless" plastic surgery will soon augment your perceived flaws in the mirror of your phone screen. But instead of encouraging people to permanently alter their bodies, maybe these face filters will quell the demand for cosmetic surgery. Why make a real-world change when you can endlessly experiment with virtual ones?

Most Snap filters are used in selfie mode, but your phone's other camera—the one that points out at the world—is being augmented, too, with more than just Pokémon Go. Brands like LEGO now have product packaging that pops up animated time-lapse "builder experiences" when you point your phone at the box. Soon many more product companies will use SuperSight to make their packaging "work harder." They'll animate scenarios, spark your imagination, explain their product line, cross-sell, and gain your loyalty.

In research and innovation labs today, every smartphone maker, social media giant, game studio, and wireless carrier is racing to innovate, demonstrate, patent, and "own" this next platform, or at least key parts of the software/hardware stack. A new landscape of companies is emerging to create compelling software applications, wearable hardware, and business models for the coming wave of SuperSight. Tech firms are building on one another's software platforms and tools like a giant layer cake, finding applications and limitations, then new approaches and solutions. Embodied multimodal interaction design, for example, is the practice of studying the right combination of free-hand gestures (like pinching, blooming, and swiping), head gestures, voiced commands, and staring at things to tell a system what you want to do—work that will determine the vocabulary we use to interact with these digital layers given what we see.

The market for this next wave of computing is vast. Even conservative financial analysts project a compound annual growth rate for the AR market of 46.6% from 2019 to 2024, growing to $72.7 billion in revenue by 2024, and over $300

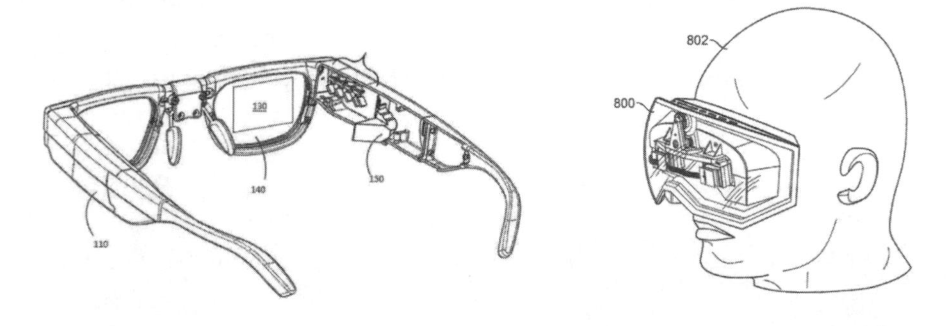

Every large platform is working on SuperSight. Evidence of this is in the increasing pace of patents from companies like Google, Apple, and Samsung.

billion by 2028. The most significant market driver is rising investments by nearly every tech giant: Amazon, Google, Apple, IBM, Microsoft, Intel, Comcast, Qualcomm, and Facebook in the US; Baidu, Alibaba, and Tencent in China; Samsung and LG in South Korea; and Sony, Canon, and Panasonic in Japan. Beating many of these giants to the market are agile startups such as nReal, which offers impressively bright and lightweight AR glasses for $500 that tether to your phone to leverage its computation and connectivity. Even more ambitious companies are experimenting with SuperSight contact lenses; Mojo Vision already has working prototypes.

Every decade, a new technology arrives that aggregates many innovations and creates a step-change in how we live and work. A decade ago, only the most prescient futurists saw the rising opportunity for smartphones (multi-touch screens + cameras + telephony), drones (sensors + flight-control algorithms + long-range wireless), voice-based digital assistants (natural language processing + voice recognition + cloud services), doorbell cameras, or internet-connected thermostats.

SuperSight is this decade's convergence technology. It inherits the last thirty-plus years of enabling technologies like machine learning, computer vision, wearables, edge computing, 5G wireless, deep personalization, affective computing, and new interaction paradigms like gesture and voice—packaged in the familiar wear-all-day form of glasses. As these component technologies mature, miniaturize, and converge in smart glasses, their impact will reverberate across every sphere of life, changing how we interact with information and each other. Smart glasses will be as normal and ubiquitous as smartphones are today.

In each chapter of this book, I examine a new facet of this inevitable augmented world: from how we connect, eat, shop, and collaborate, to the future of learning and imagination. I hope to, if successful, help you anticipate this shift, or even spark an idea for a product or company.

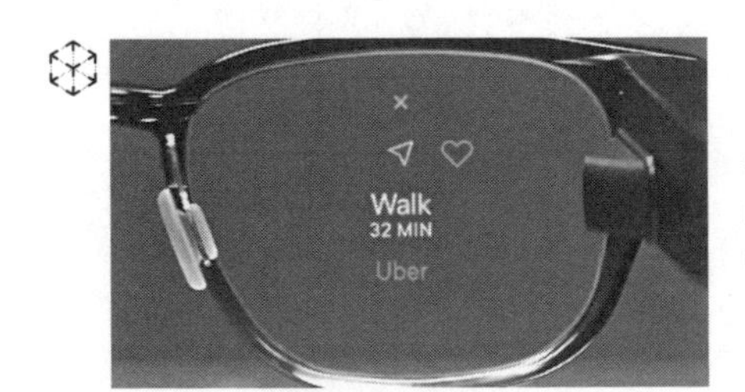

Glasses by North, now a part of Google, provide navigation, a personal teleprompter, and your Uber ETAs.

Admittedly, it's hard to forecast the slope of the adoption curve for SuperSight-enabled glasses. How swiftly will they expand beyond work settings? I expect it will initially follow an Apple Watch trajectory, since they'll be expensive ($500–1,200), discretionary, and require a certain amount of care and feeding (like charging nightly). Following this playbook, they'll enter the cultural zeitgeist as fashion and self-branding statements first, and functional assistants second. But just as mobile devices, voice interfaces, and smart watches have done during the 2010s, SuperSight will become deeply woven into culture. Smart glasses won't just change *what* we see; their ability to compose reality will alter *how* humans see. Because they are best regarded as all-day prosthetics, they won't be seen as a separate "optional" tool. Psychologically, wearables merge with the body, and like watches, shoes, helmets, or headlamps, represent an evolution of our human abilities—a part of us.

Smart glasses provide an unfair information advantage to those wearing them; it's like using Google to play Scrabble. They grant the ability to recall, visualize, and analyze—to see the complexity and interconnectedness of the world in deeper, more immediate ways. They enable us to expand our perception across the electromagnetic spectrum into the shorter infrared and longer ultraviolet spectra as some animals can, to see depth like a bat through fog and darkness, and to see into engines and buildings and through bodies of water as if we had X-ray vision. Furthermore, these spatial projections allow us to see over elastic *time scales,* and at broader *physical scales,* than we can fathom today. Granted these superpowers by SuperSight, our perceptual system will continue to evolve.

SuperSight systems are also miniaturizing rapidly. Now an entire optical system—including a microcontroller, battery, and antenna—can be packaged into ordinary glasses, a shirt button, or a swallowable pill to detect colon polyps as an alternative to endoscopy. Now that's fun watching for the whole family!

Snap's Spectacles are primarily video recorders that let you upload whatever you see directly into your feed. (Plus, they're really stylish.)

In the healthcare chapter you'll learn how computer vision algorithms have surpassed human doctors for many diagnoses. But you'll also learn what this means for the future of work, not only for those in the medical field but for anyone who has a job that might be empowered or threatened by visual AI. SuperSight's innovations are, after all, disruptive. Even the most beneficial technologies have downsides.

Evolutionary advancement inevitably comes with trade-offs and losses. Biology teaches this vividly when we examine visual systems across the animal kingdom.

Eighty percent of the human eye's receptor cells are rods, which respond to changes in brightness; the remainder are cones, which react to color. Cats' eyes, however, are 96% rods; they aren't just pretending not to appreciate your new pillows. The ratio of rods to cones in the eye follows a clear pattern: meat eaters tend to see less color because their vision systems evolved to prioritize speed, light, and depth of field to track prey. By contrast, day-foraging humans and herbivores evolved to distinguish tasty purple berries from adjacent pink poisonous cousins. Owls' eyes are so optimized for aperture that evolution eliminated their ability to rotate them—an owl can see far into pitch-black darkness, but to glance to the side, it must turn its whole head. A similar trade-off occurs underwater: natural selection traded binocular vision in fish for a very wide field of view. Seeing prey creep up from behind is more important than perceiving their world in 3D, as most of the animal kingdom does (they'd be terrible at Jenga).

The same optimization trade-offs are at play in technology—and the more powerful the tool, the more consequential its trade-off. In China, the government is using SuperSight to make roads safer by equipping city buses with AI-powered cameras that allow traffic cops to spot reckless driving. While safer transportation seems positive, in this case it jeopardizes privacy. AI-equipped drones might be able to help us monitor construction site safety and prevent poachers from killing rhinos, but authoritarian governments use the same technology to control dissent by surveilling and profiling civilians. As my MIT colleague Kevin Slavin has quipped, "It's a bright future if you're an algorithm." But if you're a human, especially one living in a state without strong privacy regulations, perhaps not so much.

Almost every computer vision application that I raise herein is, at turns, fascinating and frightening. Predicting the ramifications of these technologies and their consequences is both worthwhile and devilishly complex. But just because the impact questions are difficult and slippery doesn't mean that technology will bring dystopia. Like moveable type, the telegraph, electric lights, penicillin, the automobile, or CRISPR, SuperSight is a monumental invention both promising and perilous for society. And the more thoroughly and vividly we imagine augmented futures, the more likely we'll be able to integrate spatial computing ethically, regulate its downsides, and use it as a force for good. I hope I can deepen your understanding and invite you into this important conversation.

As we work through the consequences of SuperSight in the coming chapters, I highlight six distinct *hazards of SuperSight* and suggest measures we should take to moderate or neutralize each. Each represents one or more sets of interests in tension: profitable business models versus human mental health, or frictionless personalization versus strong privacy protections:

⚠ SOCIAL INSULATION

When we're all walking around in our own self-contained, personalized realities every day, human connection and our sense of community may suffer. When we can all choose to add whatever graphics and skins we desire to the world around us, it'll be literally harder to see eye to eye. Computer vision could trap us in our own personal views of the world, impeding our ability to understand and empathize with others.

⚠ STATE OF SURVEILLANCE

Cameras, now installed in everything from school hallways to household doorbells, will become increasingly pervasive through being embedded in the frames of our glasses. These cameras are getting more capable, attuned, and sentient. As this data is stitched together to provide useful services like personalization, it will also provide governments and private companies unprecedented and incriminating access to what we like and buy, where we go, what we do, and with whom.

⚠ COGNITIVE CRUTCHES

Assistive technologies like GPS often lead us to lose skills that we no longer practice, like map reading, handwriting, or direction finding. Spatial computing allows us to turn the whole world into a classroom, but when we all have personal agents coaching us as we play tennis, repair our homes, and go on first dates, we could become over-reliant on them.

⚠ PERVASIVE PERSUASION

We are already accustomed to trading personal data for free digital services (hello, Google and Facebook). In the age of computer vision, companies and brands will not only be able to see our search history and events calendar—they'll be able to see what *we* see, too. This means they might start influencing our behavior and purchases as never before.

⚠ TRAINING BIAS

To rely on computer vision's judgment—be it about a medical diagnosis or a choice to speed through a yellow light—we must trust our lives to autonomous systems, often without a clear sense of their accuracy or training methods. There are already huge biases in the datasets that SuperSight learns from, especially when it comes to race and gender.

⚠ SUPERSIGHT FOR SOME

In the development of any technology, societal inequality becomes entrenched early. Will we create even more of a digital caste system as SuperSight progresses?

Suddenly feeling ambivalent about computer vision, face recognition, AI, and the Internet of Everything? You're not the only one. I struggle with the seemingly inevitable and unsettling consequences of these technologies, even as I've helped develop some of their most innovative applications.

As an entrepreneur and futurist, I'm captivated by opportunities to redesign and reinvent with new materials and nascent technologies. I studied physics and fine art as an undergrad—two disciplines trying to understand and capture light—and have been an avid photographer since I was a preteen. When I was eleven, I built a darkroom in the basement of my parents' house in Madison, Wisconsin, and have been rolling my own black-and-white film, burning, dodging, and inhaling fixer ever since.

In college, I mounted photo exhibitions around campus. It was always thrilling to show my work and receive feedback from others. However, outside of St. Olaf College and my parents, no one could see my photos. That was why, in the 1990s, I became so excited by the internet. I loved the idea of sharing my photos *online*, where anyone around the world could see them, not just folks studying in my college library. Back then, though, the internet was mostly meant for text and static images. A fellow MIT grad student, Neil Mayle, and I patented the process of uploading photos through a web browser to invent the first online photo-sharing service, then raised money and started a company in 1996 to host online photo albums. But we were too early for digital photography: only pros could afford digital cameras, home modems were 56k baud, and it cost an additional $30 to get thirty-six photos digitized onto a CD from the corner pharmacy's Kodak film developer. We never envisioned the simple, endless photo scroll that has become the standard structure for Facebook and Instagram, or that in two decades we'd be walking around with all-you-can-eat cell plans and multi-megapixel cameras in our pockets, sharing endless selfies and videos of cats. (Should have kept that company, alas!)

Instead of continuing to pursue photo sharing, my thriving product design firm, Interactive Factory, went on to help build robotic toys like LEGO Mindstorms and music-learning experiences like *Guitar Hero*. We created simulations to teach physics and computer science, language learning software, and interactive science museum exhibits all over the country.

A virtual race experience adds motivation to rowing at home.

In 2000, I was leading a multidisciplinary innovation group at Viant. We had just gone public, so I convinced the CEO to become an academic sponsor at the MIT Media Lab, where I'd been a student. As part of one of these collaborative projects, I became obsessed with the opportunity for subtle, *glanceable* peripheral information displays, so I started a company, backed by the founder of the Media Lab and Hiroshi Ishii, one of my mentors there, called Ambient Devices. There, "enchantment" was my organizing metaphor—a way to reimagine the world as a place where every object was connected, smarter, and animate. My first book, *Enchanted Objects,* explored the coming world of the Internet of Everything from an inventor's perspective. (Now, thanks to the next wave of immersive technology and computer vision, we no longer need to embed sensors and Wi-Fi chips into mirrors, tables, kitchen appliances, LEGOS, and lamps to personalize and add services to these objects. Instead, we'll use SuperSight to combine the digital and physical, so we can see the spaces and objects and people around us entirely anew. SuperSight allows us to do the enchanting from the outside.)

In 2015, inspired by the opportunity to use AI to enable social shopping experiences through social media photos, I got together with Neil Mayle (previously my co-founder at the photo-sharing company) and another Media Lab colleague, Joshua Wachman (previously my co-founder at a healthcare company), to create Ditto. When we launched the company, 300 million photos were being posted on social media every day. These pictures contained things that inspired people and captured their interest, but the contents of these photos were invisible to the hyperlinks of the web—they were "unstructured data." We trained a cloud-based algorithmic brain to identify thousands of brands, objects, fabric patterns, and contexts of use, then link those objects or experiences to their sources: ecommerce sites, travel agencies, restaurants, sports-ticket sellers, recipes, eBay, the local pound. The goal: when you were scrolling through Facebook photos posted by your friends and thought, *Nice backpack/shoes/standup paddleboard/lamp/ski vacation/Red Sox game/blueberry pie/pug; I wonder where I might find something similar?*, you could "Ditto that."

After Ditto sold, I was recruited by the glasses company Warby Parker. As VP of vision technology, I used my background in physics, computer vision, and digital product development to create online vision testing. We patented the process of positioning people at a precise distance from an on-screen eye chart using computer vision running on their phone to accurately perform at-home eye exams, creating an online test that is now accessible and convenient for the billions of people who need corrective lenses. Then we paired that test with a virtual try-on tool that uses AR to measure your pupillary distance and the contours of your face to recommend and let you visualize which frames look best. During my time there, we also researched and prototyped a new generation of glasses that embed technologies like hearing aids, focus-tunable lenses, augmented visual information, and sensors that read brainwaves to infer emotion and focus.

At the product design firm IDEO and now at EPAM Continuum, I work with teams on speculative-futures projects. You'll read about many of these in this book: the future of toys for Fisher-Price, a streamlined experience for restaurant takeout, automated home-cleaning robots, and more. The process starts with deep customer research to discover unmet needs, wishes, and mental models (the metaphors people use to think about things). The craft is then to synthesize these insights into opportunities for businesses, make prototypes, and see what resonates. These future-casting projects live at the intersection of people-insights, tech disruption, and business model invention. It's thrilling to learn and speculate, precarious and often frustrating to work with brittle new technologies, and often humbling to see how people react to your beloved prototypes.

Throughout all of this, I have been teaching ambient computing at the MIT Media Lab, lecturing at other research labs, keynoting industry conferences, and advising startups as they build their product roadmaps. And SuperSight has been at the center of all of it.

I have been fortunate to witness firsthand the profound changes in products and business models that technology inflection points have triggered. As a faculty member at MIT, a startup investor, and a consultant to companies working in areas like fashion, healthcare, urban design, and architecture, I've become practiced at reading patterns across industries and forecasting the impact of the coming digital waves. My goal is to help you learn some of those same clairvoyance and foresight skills.

Over the next nine chapters, we'll unpack the technologies that make SuperSight possible and examine the implications of the next stage of our visual evolution. In the first chapters, we start with the ways SuperSight will change our individual experiences and interactions, then gradually zoom out to society-scale issues like food, education, work, and health, before finally turning to the area of most profound impact: the ability to transform our collective imagination and motivate change.

In each chapter, you'll meet entrepreneurs and scientists pioneering SuperSight technologies and services: Salvador Nissi Vilcovsky inventing magic mirrors for high-end retail; Jenny Boutin pairing micro-gardens with celebrity chefs; the inventors of the Smoke Diving helmet that helps firefighters navigate with X-ray vision. I'll also introduce you to hardware and software companies that are finding ways to compete with the "big five"—Google, Amazon, Facebook, Microsoft, and Apple—all of which are investing billions to own the next platform for how we synthesize and experience reality.

As we go, I'll explain the technology and key algorithms that enable SuperSight, and introduce frameworks to understand this coming world. More critically, I'll look at important questions about impacts and secondary effects of SuperSight to help you imagine the products and services of tomorrow.

We are living through an era that requires new eyes. We must make essential issues more legible: the consequences of climate change, the pervasiveness of inequality, lack of access to education and healthcare. What improvements might we be able to effect with evolved sight? And is it the night vision of owls, the distance acuity of a falcon, the peripheral vision of fish that we require—or do we need to see something else entirely? I will argue that, among myriad other practical applications, we most urgently need SuperSight to see the future, for the sake of our own health and the preservation of the planet.

My grandfather, an architect, would reflexively grab a roll of tracing paper whenever he wanted to quickly express a visual idea. He would use it to sketch over a photograph of an existing building or landscape, then lay another piece of tracing paper over that one and change a few lines to draft an alternative. SuperSight will be our future tracing paper: the fastest way to prototype and envision personal, systemic, and city-scale change.

The future is watching us. Let's use SuperSight to envision the best one we can.

One last thing: let's innovate a bit on the form of a book itself. Of course, I wanted to make this a full-color, 17" × 20" coffee-table book, but you probably wouldn't have purchased that for $150 . . . at least that's what my agent said. So, instead, you can use your smartphone to view color photos, videos, and animated diagrams when you see this glyph:

First, download the SuperSight app by pointing your camera here:

Open the app and point your phone at the page to reveal the color, animated, and voiced information . . . then, use the app to share these and your reactions if you'd like. Try it now on this thumbnail for a message from me:

If you want to just read without a phone in your hand, I applaud that decision, too. All of the color images, animations, and videos are also available on SuperSight.world.

Part 1

Human Scale

This first section explores how advances in computer vision will let us understand the world around us differently. It will change not only what we see and experience, but also how we are seen. This new visual feedback loop will transform how we learn sports and hobbies, interact with others, and make everyday decisions. These sentient services will also have profound secondary effects on personal safety, privacy, equity, and our physical and mental health.

What if you knew everyone's name?

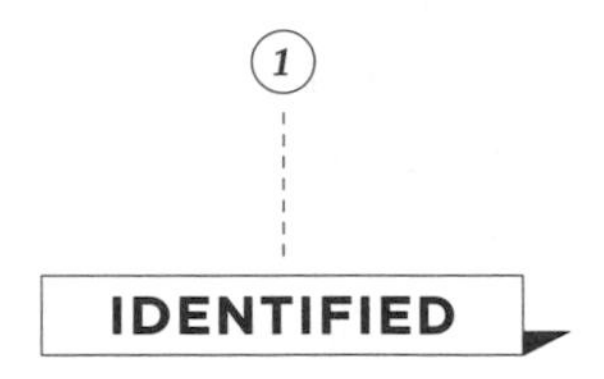

About human interaction and the consequences of a labeled world

1.0 A walk in the woods

metadata *augmenting nature*

Let's go for a walk in the woods together. It's springtime, and the forest is coming alive with sounds and smells. Curious to learn more about all the things budding around us, we both enable a nature app in our glasses that adds a virtual layer to the real world. We first notice little nameplates on nearby trees: "Red Pine, 1918"; "White Oak, 1775." Thirty feet up, we spot holes in the tree trunk and see a faint image of a pileated woodpecker looking for a snack. We can discern that he's digitally projected into the scene because the woodpecker's form is translucent. We also see a simulation through the tree of a large carpenter ant nest that he's munching on.

When we walk past a rock outcropping, we learn that it's a history sandwich: the topmost layers are sandstone, the sedimentation of an old lake bed; then metamorphic layers caused by millions of years of pressure; and finally igneous rock formed by cooling magma. The narration on rock stratum is provided by the Danish scientist Nicolas Steno (love the accent), who first used fossils to establish the law of superposition behind the creation of these geological timelines. A virtual rope keeps us at a safe distance from the outcropping's forty-foot drop lest we get too engaged with the ad hoc geology lesson.

Now we walk out of the woods to the edge of a meadow. A red-tailed hawk, hundreds of feet above us, is riding the wind currents. For a few seconds, our perspective flies up to hers, where we see hundreds of iridescent trails on the ground, illuminated in our field of view—traces of reflective urine, visible to birds of prey that see on the ultraviolet spectrum, that show the paths of mice and voles. Next we come upon some tracks in the soft earth. Based on the pad print and spacing, we see the red fox that created these prints projected in a slow-motion full gallop, perhaps pursuing the same snack as the hawk. On a nearby bush, we observe all the birds that feast on its berries, then return the favor by distributing seeds in their dung. Curious, I ask to learn more about prairie ecology. The scene dramatically changes as smoke obscures the sun, and a wall of simulated flames thunders past me to illustrate the upcoming prairie burn. While it seems destructive, I learn that a routine of planned fires is good land management and provides a fresh start for low-growing wildflowers to flourish this summer.

So how does this all work? How do we take all of Wikipedia and spatialize it into a walk in the woods?

This first chapter is about the first "gift" of SuperSight: the ability to recognize and name what we see. Naming is the first kind of knowing. Growing up, our parents and teachers label things for us. I remember childhood walks in the woods with my parents, who would invariably name the things we passed: oak, elm, cardinal, sparrow, moss, tent caterpillar. Picture books such as Richard Scarry's *Busy Town Busy People* served the same purpose, cataloging and naming thousands of objects in scenes and wonderful cutaways of ships and city streets. These labels satisfy an innate biological yearning to identify our surroundings. Nouns are the first things humans learn; they are the essential blocks with which we build more complex ideas, opinions, and arguments. Red pine, sandstone, vole urine—nouns help us categorize, organize, and therefore understand the world around us. A new digital layer of information and holograms will accelerate this process.

Labels and captions, in whatever language, will be the first app you enable in your smart glasses. When I was first learning French in high school, I found a French sticker book and labeled everything in our home: *bureau, toilette, bougie, livre, sel, poivre, mère, père, raquette de tennis, bicyclette, ordinateur*. Because the labels were visible as I encountered these objects throughout the day, my recall improved.

Look around you now. What objects "want" a link to a definition, a history, a voice? Of course you can type anything into Google, but SuperSight allows an immediacy to information, the ability to access knowledge simply by letting your gaze linger on an object.

In the room around me right now, I want to see not just the names of artifacts, but metadata (data about data), stories, and voices. When did Grandpa get that old antique sideboard? How did he afford it as a new immigrant from Germany? Who designed it, crafted it, and with what tools?

Metadata about the world won't only give objects voices. Spatially located data also has the potential to keep us safe. As my kids ride their bikes around Boston, the color of the pavement will change to reflect risk. They will be more likely to slow down and look twice as they approach orange-glowing intersections. In the kitchen, the same "risk auras" will surround knives, graters, and hot pans recently pulled out of the oven. These auras will also highlight slippery spots after a winter storm, or just poorly lit stairs—especially useful for seniors, for whom falls are so common and debilitating.

While most of what we envision for augmented reality and spatial computing is additive, SuperSight can also obscure, blur, or *diminish.* When grocery shopping, for example, my gluten-free mother doesn't want to see products that will make her sick (and maybe not the predictably irresistible chocolate, either). I would rather not see billboards when driving, and most wayfinding and warning signs could be eliminated when I know where I'm going. Please replace parked cars with bushes.

SuperSight will induce a kind of collective synesthesia, in which we'll be capable of perceiving and seeing more—the past, the future, emotion, intent, value, risk—as visual halos. For designers and entrepreneurs especially, this is an exciting new superpower to contemplate and enable, offering myriad new opportunities and consequences, both incredible and troubling.

But before we consider the downsides and risks, it's essential to cover how we teach computers to see and the limits of this capability.

1.1 Training a computer to perceive

deep neural networks | *training data* | *precision and sensitivity*

At first, SuperSight seems like magic. Google Photos knows who is in your photos and recognizes babies and dogs better than you. Pinterest finds similar fabric patterns to an image you uploaded from all over the web. And cars can read the speed limit off of road signs. But it's not enchantment; it's just algorithms and a lot of training data.

The origins and ambitions of computer vision trace back to the 1960s. Back then, everyone thought it would be easy to convert pixels to understanding, but as it turned out, teaching computers to see is nontrivial (a physicist's phrase for *wicked hard*). Conversely, text indexing and search are relatively straightforward, but it takes sophisticated and computationally hungry algorithms to figure out what is in an image or video. How can a computer know what objects are in a photograph? And *where* those objects are in the photograph? *Object recognition* and *scene classification* are our first building blocks of SuperSight—the easy-to-read top line on the vision test.

Imagine again that you're taking a walk in the woods and a bird flies by. How do data scientists train an app on your phone to differentiate between a heron and a bald eagle? It starts with "training data": thousands of photos of labeled herons or eagles in various poses on various backgrounds (these are called *true positives*). To train a robust algorithm to recognize an eagle in all their poses and forms, you'll need images of eagles in flight, nesting, standing, and swooping; young and old, male and female, bald and golden, and so on. You'll also need thousands of photos that have *no eagles* present but that include things that are typically found in the bird's environment (these are called *true negatives*). This way the algorithm can learn to differentiate between an eagle and all the things found around eagles, like trees or flags with eagle logos on them.

The iNaturalist app has done this to help people identify thousands of plants, animals, bugs, and flora. My daughter and I recently climbed the three hours to Mount Lafayette, one of the most picturesque hikes in New Hampshire's White Mountains. As we sat exhausted and euphoric on a rock at the summit, I noticed a lovely tiny flower somehow eking out an existence at 4,000 feet. Curious, I took the following photo in iNaturalist and was delighted to find out its name,

Diapensia, and some details: "Sometimes called the pin-cushion plant, it is also found mainly in the Himalaya and in arctic habitats."

A high-mountain flower, *Diapensia*, that I photographed on a recent hiking trip and identified using iNaturalist's deep learning algorithms.

iNaturalist launched in 2008, trained on a set of hand-labeled photos from ImageNet, an academic project started at the Stanford Vision Lab by Fei-Fei Li, now Google's chief scientist of AI. This database now has more than 14 million hand-labeled images with more than 20,000 labels (thank you, grad students). Professor Li created this massive database of labeled images to judge the quality of image recognition algorithms in the Large Scale Visual Recognition Challenge. In 2010, a winning algorithm boasted accuracy of scores by around 70%. Fast-forward to today, and winners score 95%+ accuracy, consistently performing better than humans asked to do the same task. Could you pick out ninety dog breeds with more than 95% accuracy? I couldn't.

iNaturalist automatically labels bugs, bats, and herons in photos you upload. This stream of user-contributed data increases the precision of its algorithms.

What is brilliant about iNaturalist is its structure of iterative crowd-sourced training. Graduate students weren't required to teach it the difference between a thimbleberry and a marionberry—its users do it for them. As people upload more pictures of plants and bugs, iNaturalist builds an ever-larger training dataset, which increases the diversity of shots, which in turn increases the app's power and accuracy. If the AI isn't confident about a label, then a researcher steps in to confirm the guess or manually add the correct one. New data is then fed back to retrain the network, making it even more precise. (Computer vision systems love a big self-feeding loop.) By being exposed to more and more trained data, iNaturalist now has an accuracy rate on most flora and fauna of over 98%.

Crowd-sourced results from iNaturalist are geotagged to show a useful layer of metadata: Where in Brookline should I go to find wildlife?

The most advanced computer vision algorithms used now by nearly all of these applications are deep convolutional neural networks. They're called *deep* because there are multiple layers, each tuned to find different-scale features in an image. The bottom layers focus on fine details like patterns; other layers detect gross features like shape. As each of the nodes in this network is activated, a unique vector (in essence, an image's signature) is calculated and matched with a library of other vectors. When there is a high-confidence match, then the image is properly labeled: a bat, a bug, an eagle.

How good are these detection algorithms? The answer has two facets: sensitivity and precision. Imagine you're a radar receiver operator in 1940. You are deployed in a new, top-secret station on the coast of Dover installed to identify incoming German planes with something called radar. It's your job to stay awake all night and listen for incoming static, then decide whether it's coming from a low-flying plane (in which case you sound the alarm) or a flock of seagulls (in which case you let people sleep). Your only controller is a large sensitivity dial. Dial up the sensitivity, and you're more likely to detect planes (true positives), but you'll also detect more geese (false positives). Dial it down, and you'll detect fewer geese, but you might also miss a plane (true negatives). It's not obvious where to set this dial! The system is either crying wolf or letting the wolf in undetected.

This is also how we measure the quality of an algorithm: in sensitivity and precision. *Sensitivity* asks what percent of all plane-like objects are detected, and *precision* asks, of the objects detected, what percent are actually planes (versus birds or whatever).

I've never met a perfect algorithm. There are always false positives (the algorithm says it's nothing, but it *was* a plane) and false negatives (the algorithm says *plane*, but it was just a goose).

So, like the radar operator, you have to choose whether to privilege sensitivity or precision. Most medical tests prefer sensitivity: mammograms crank up the sensitivity dial because it's better to have a false positive (you spot a bird that might be a plane or a shadow that might be a lesion) versus a false negative (where you miss both the plane and the bird, and early breast cancer).

Modern computer vision systems always return a confidence number with any prediction to aid decisions about which data to act upon. For example, you could be 76% certain that it was a plane, or 12% certain that it was a bird (however, we can never be 100% sure it wasn't Superman). Using this confidence number, a data scientist can decide a cutoff, set a criterion for throwing out low-confidence results . . . and just let everyone sleep.

Where you set that cutoff depends on the application. If you're trying to spot a raccoon before it gets into your trash, you'd want it to sound an alarm at maybe 80% confidence; the worst thing that happens if you miss a few raccoons is a mess you have to clean up. Comparatively, you want higher confidence for catching spy planes or identifying a lump in your armpit. In healthcare especially, the cost of missing something is high, so you're willing to have more false positives.

For SuperSight to function, it first needs to be confident about what it's seeing. Training it to see underlies every single other thing we talk about in the book. Now we can turn to what SuperSight might do with this perceptual information.

1.2 Knowing the name for everything—and everyone

augmented conversational scaffolding | *least-common commonalities*

I once met an army colonel who confessed that his primary job in Washington is to memorize key facts about hundreds of people and deliver this information under his breath to his boss, seconds before it's needed. He's a walking contextual database with millisecond recall. Before going to a networking event, he'll spend the day with its guest list and his personal "face book." That way he can walk around behind a general and whisper tips to him about the people he meets: names of spouses, kids, and dogs; sports interests; and, importantly, what the general's "ask" is of each person to advance their department's goals.

We all could use this guy—a computer vision–triggered, just-in-time relational database whispering in our ear. Such assistive agents will soon be a standard feature in SuperSight glasses.

Your glasses will have a tiny front-facing camera with a neural network trained to recognize faces, then whisper names to you as you approach, or display them subtly as subtitles under others' chins or on their foreheads, so you never lose eye contact. The system doesn't need to recognize and label every face in the world, only those of people you've met, or would like to meet. Facebook and LinkedIn have the mug shots, the social graph of who knows who, plus key content associated with each person: expertise, interests, work history, reputation, and the like. Of course, SuperSight won't do everything for you; your glasses won't capture how you can help others, or how they might assist you. You still need conversational skills and élan.

Beyond placing names on foreheads, our SuperSight glasses will reveal topics that an exquisite party host would introduce—the least-common commonalities. These shared experiences and interests will bind people together and create quick rapport: you both own a Labradoodle, love Drake, visited Cuba, grew up in a small town. Many people will opt in to share overlaps in Twitter content, podcast or streaming media consumption, recent travel, or current conversational topics; others with higher levels of privacy settings might look like a blank page to you, without even a name for a cue.

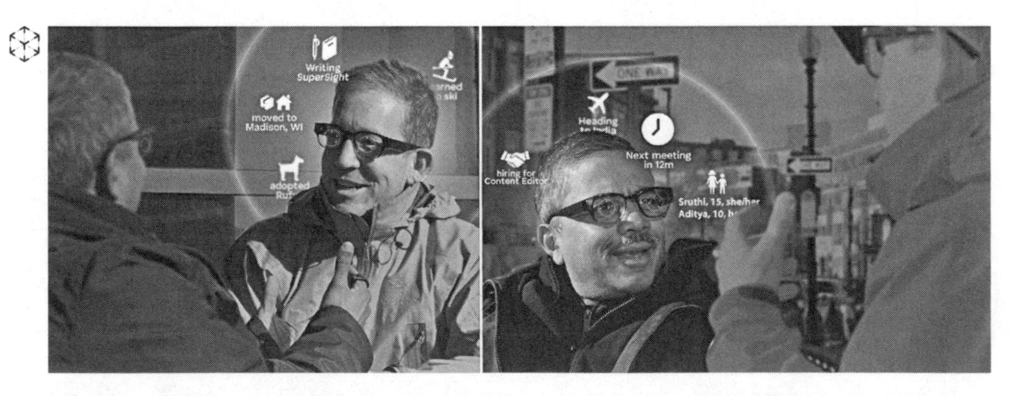

Metadata clouds of conversational cues will surround other people like hieroglyphics.

In this way, an aura of metadata, summed up in glanceable symbols, will surround the people with whom we engage. This swarm of hieroglyphics will suggest topics that are most likely to increase intimacy and rapport, and that help you get "conversational jobs" done, like a checklist of *don't forget to ask them about . . .* or *be sure to ask them for . . .* To illustrate this, I met a friend, Hari Nair, in Harvard Square one morning while wearing AR glasses from North, a company since acquired by Google (see the previous photo). We have a long history of working together on Internet of Things projects, both when he led the innovation team at Kimberly-Clark, the world leader in diapers, and more recently at cleaning and home products giant Procter & Gamble. The cloud of icons that we prototyped swirling above his head that morning included emotional-intelligence cues like asking him about his children and being aware of his busy schedule, acting on interest overlaps like his upcoming trip to see his mom in India (I've always wanted to go there and would like his advice), and reminding me to ask about hiring an editor for an article we are working on together. His prototype glasses saw different icons orbiting my head: they cued him to discuss a case study on robotics for my book (see chapter four about kitchen-cleaning robots!).

Our cloud-connected SuperSight glasses *could* show us anything, and could be incredibly distracting and overwhelming, like smartphone apps and web browsers are today. But since we are designing a see-through experience that knows where you are and what you are looking at, the best designs will prioritize contextually anchored information. Instead of overwhelming you, this metadata should be filtered down and curated based on its spatial and contextual relevance, informed by your goals and interests—making it more useful, valuable, and actionable. Most of the time, smart glasses should show nothing at all.

To test this "contextual relevance" premise, I programmed a fun hack for a party last year. I put on MagicLeap AR glasses that have a camera on the front and a discrete speaker near my ears, then fed the stream from the camera into an image classifier algorithm so that any object in my field of view was labeled. My program then took those labels and fed them into a search engine for jokes. Then, when others weren't talking, it whispered a joke into my ear. When I walked up to someone holding a guitar, it recognized the instrument in their hands with computer vision and fed me this: "What did the violin say to the sad guitar? Don't fret." I delivered the joke to the delight of the strummer and groans of everyone standing around. "Don't worry," I added, "I'm here all night." This simple experiment shows how SuperSight glasses will make us appear smarter in conversation—or at least turn us into corny comedians.

Many will enjoy the new efficiency of knowing more as we move through the world, but some will be concerned that whispering agents might exert too much control in conversations, or limit the messy but interesting serendipity that defines a good discussion. Recommendation engines could lead conversations to feel more transactional: "That woman in the blue blazer is hiring for an associate position that you're qualified for" or "You and David both watched that Netflix true-crime documentary last night, so ask him about it." We may begin avoiding interactions that aren't "advised," and lose the ability to stumble onto a totally wonderfully odd topic.

That isn't the only risk presented by SuperSight's information-rich future. In his 2019 novel *Fall*, Neal Stephenson imagines a similar future where everyone wears an AR screen that flips down like a visor. To curate this "feed"—to see the star rating for restaurants and label the people approaching you on the street—you hire an "editor." This is either a real person or an AI—depending on what you can afford—that helps parse the deluge of information spewed at you and protect you from harmful content. However, the novel notes, "few people were rich enough to literally employ a person whose sole job was to filter incoming and outgoing information."

One of the flip sides of seemingly useful AR classification systems is the shadow log: scoring, branding, and labeling undesirable things—and people. There may be "hazard ratings" over the heads of some people we encounter: glowing red might indicate a history of convictions or a less-than-five-star Airbnb rating. Yet this may cost us our capacity for forgiveness and absolution, and people may never receive a clean slate from past misdeeds.

Or, rather than cluttering our visual field with orange glowing things, we may choose to visually filter, blur, or erase sets of undesirables from our view—not just awkward dates or exes, but those with dissimilar views or values, or those that algorithms deem to be less attention worthy. This will be catastrophic for general empathy and drive further social divides, imposing an even more homogenized, isolationist worldview upon us. These "filter bubbles" may leave us unable to spot and appreciate, for example, systemic racism and inequality: who we see could determine what we believe.

⚠ HAZARD: SOCIAL INSULATION

Social insulation, a hazard of SuperSight. We may all see a different view of the world.

SuperSight risks cocooning each of us in our own private view of the world. When we inevitably choose to see layers of information different from one another's, it will become increasingly challenging to connect with others around shared experiences, or even understand others. He may use his glasses to see practical information like weather forecasts, wayfinding, and plans, while she has a penchant for playful monsters, historical fiction, and imaginary world-layers.

We've already seen how filter bubbles in our social media feeds divide us socially and politically. As immersive experiences become increasingly personalized, these bubbles risk becoming sealed and inescapable. This would profoundly affect our communication, sense of community, and civic practices.

A solution would be to let people quickly synchronize views with others, perhaps with a high-five or bump gesture. This would be akin—in the audio space—to sharing a headphone splitter or a Bluetooth feed so everyone is "listening to the same music" even when wearing their own headphones. We'd also want a way to swap views—to see the world through others' eyes. An easy view-toggle might spark heartier conversations with more reasons to engage, connect, and invent. Instead of creating a filter-bubble problem where we self-select information that feels familiar to us, we could just as easily subscribe to services that introduce us to something entirely different. We might experience a point of view we've never encountered or taken seriously.

In this way, the AIs could help expand, rather than diminish, our social networks and the topics we engage around. The key is for the AR industry to follow the ethos of open architectures and open standards that allow companies and individuals to configure, remix, experiment, hack, and share their version of SuperSight's reality-bending technology.

1.3 Cameras woven into the fabric of our lives—literally

lifelogging *mirror neurons*

James Bond's ingenious inventor Q introduced us to some incredible Cold War spy tech: gadgets that gave 007 the ability to breathe underwater, levitate to safety with a quick zip line, generate a smokescreen, or eject a villain from a moving vehicle—and spy cameras tiny enough to embed in carnations pinned on the lapel of Bond's tuxedo. It may be illegal to own an arsenic-tipped umbrella or a ballpoint pen that shoots poison darts, but today, bespoke camera tech is embedded not only in SuperSight glasses, but in everything from your doorbell to a pill that can spy on your colon.

Cameras owe their rapid evolutionary progress to the smartphone: its massive economies of scope and scale, and the industry's desperation for differentiation. Camera innovation has experienced incredible investment from research groups at Apple, Google, Motorola, Nokia, Samsung, Sony, and more. Factories pump out billions of increasingly capable cameras every year, driving the COGS (cost of goods sold) for a 5-megapixel low-light camera to less than $5.

Now, phone cameras are mini-computers, too, with dedicated onboard silicon and neural networks. These chipsets analyze the type of shot (e.g., landscape) and recognize specific objects (human faces and smiles) even before sending an image to the cloud (the industry's optimistic name for internet-connected offsite data centers that free you from having to store information on your phone or laptop), where even more algorithms and judgment await. This means that cameras can now reason about images to optimize exposure, or take multiple shots of a wedding party and select the one where no one is blinking. They run neural networks right inside the camera to detect objects and scenes, too. For example, your smartphone can recognize a macro food photo, a portrait of your dog, or a snap of a sprinting athlete. Then, knowing the content and context, it can automatically pick the right exposure and shutter speed. And all of this happens nearly instantaneously in a pocket-sized device.

Performing all of this processing inside the camera without needing to send data to the internet is called *edge computing,* because the computation occurs at the "edge" of the network, not in some Iceland server farm cooled by glacier runoff. There are a few reasons to appreciate the advantages edge computing brings us. It works in receptionless areas like remote Nepal or inside a tunnel; the processing speed is typically very fast, saving you bandwidth costs; and you can employ privacy-by-design principles—if you don't need to

Cameras are making their way into wall screws and shirt buttons (Q, eat your heart out!).

send images to the cloud, you don't have to decide how long to store them, or who has the right to view them. Edge computing makes real-time SuperSight possible.

Now that we live in a completely camera-saturated environment, everyone can *Get Smart*. That's not just great for catching crime lords; it's also useful for understanding ourselves.

A few years ago, I went down a rabbit hole on eidetic memory, the rare ability to hold a visual image in the mind's eye with such clarity that it can be perfectly recalled. To experiment with the ramifications of this superpower, I started wearing a tiny camera clipped to my shirt that took a photo every ten seconds (as I biked to work, spoke at a conference, taught at MIT) and then uploaded it to the cloud. At the end of the day I scanned these little flipbooks to see who I interacted with, how much time I spent in front of a screen, how much coffee I drank, and so on. The difference between human recollection and photo-recorded evidence is somebody's future PhD thesis. I could have sworn I only snacked twice in the afternoon, or spoke with just a few people at a conference, but the evidence my lifelogging camera captured proved otherwise.

To automatically generate analytics on my days, I ran these photo flipbooks through a deep learning network for automatic tagging. I could see how many minutes I spent in conversations versus in front of a screen, how often I ate, who I spent time with and if they seemed to be speaking or listening, time outdoors, time dancing in front of the mirror (just kidding), and many more metrics. It

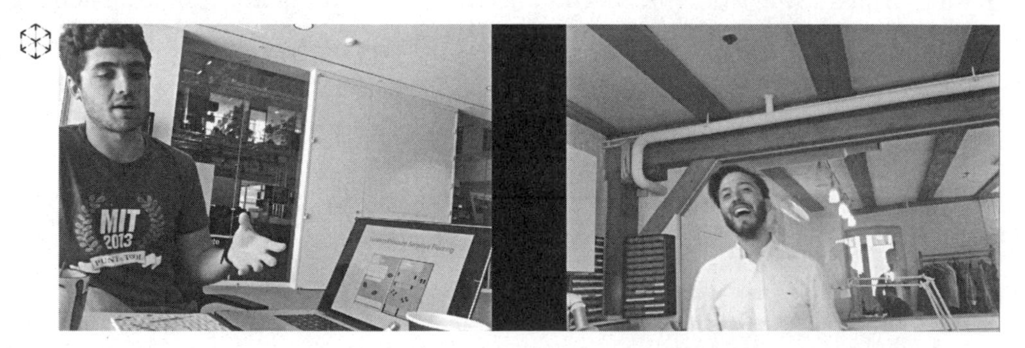

Lifelogging at MIT and in Copenhagen. Processing the stream of these images revealed surprising details about how many people I interacted with, the frequency of my snacking, and how my own mood was mirrored in their expressions.

was like a visual Fitbit on steroids, except for my personal interactions instead of my heart rate.

One of the most surprising results arose from being able to track other people's expressions as we interacted. On some days, people were smiling and laughing in front of me more often; on others, they were more serious, even glum. Was this related to the barometric pressure, time of day, or just the persistent pouty look of the French participants in this Paris business conference? (Everyone in Australia seems to smile regardless of their mood.) Using a smile detector algorithm (more on that in chapter two), I scored others' dominant facial expression as captured by pictures and looked at the data. It turned out to be a good proxy for my own mood. Psychologists claim that our mirror neurons quickly respond to the affective state of others, adjusting our speech rate and tone, and reflecting their humor or inheriting the dark clouds over their heads. It made me more aware of how my feelings affected those around me, which in turn kept my mood up throughout the day.

Inspired by a wearable computing assignment, one of my MIT students started wearing a front-facing video camera with the lens woven into his backpack strap. His experiment was to record his entire life while also wearing a galvanic skin sensor on his hand to measure his second-by-second emotions throughout the day via the salinity/conductance of his skin. He then tagged the video with this emotional track, and automated a daily highlight (or lowlight) reel. By watching each day's reel, he better understood the causes and triggers for his mental states, and who/what to avoid the next day.

Expect this kind of voluntary visual diary to be commonplace—as ubiquitous as Fitbits. Life logs will be uploaded to social networks, auto-sharing thirty-second supercuts of our high and low points of the day, year, or life. We'll also be able to scroll back through our days to show people memories: "You should have seen the look on his face!" someone might say, zooming through their reel to find it. Instead of asking your kids how their day was and receiving the usual bland response—"Fine"—you'll be able to watch a highlight reel of the moments when your kids were the most engaged, or cue up their moments of confusion to help them with the trickiest parts of their homework.

The *Black Mirror* episode "The Entire History of You" predicted the consequences of a world where most of the population can record and rewatch not only their days, but their whole lives. Fighting with your partner about what they said in an old argument? Suspect your husband was trading flirty looks with your friend across the dinner table? Just scroll back through your visual feed—or throw it onto the TV screen for everyone to see—to confirm your worst hunches.

This, of course, is the downside of being able to remember anything we want: the inability to forget.

1.4 Peepholes that peep back

home security *camouflage makeup*

Capturing your life with a body camera is interesting—but should we be allowed to stick SuperSight cameras everywhere to record what your family, students, employees, and even strangers do? Try Googling "babysitter camera," and the results will make you want to start tearing your teddy bears open. You'll see numerous plush toys with cameras embedded in their eyes—each a SuperSight-enabled Elf on the Shelf. The market for SuperSight surveillance and security is exploding.

We have always felt the need to protect ourselves. And fear sells. Security companies exploit this psychology of vulnerability by charging hundreds of dollars a year for the most basic set of sensors, such as door and window switches. If you can offer a sense of security—even if it's not real security—people will buy into it.

Amazon has jumped into this space in a big, bold, data-driven way by acquiring Ring, a company that makes doorbells equipped with cameras and Wi-Fi

Cameras in doorbells recognize friends, strangers, and delivery folks.

connections, and now an indoor drone with a camera to fly around and patrol your home. The price for this peace of mind? For Amazon, $1 billion. (For the consumer, a Ring will set you back about $100, plus $100 per year for a subscription service to save your videos on Amazon's cloud for sixty days.)

There are at least two motivations for Amazon to purchase "eyes in the home" for this astronomical price:

1. The home security market, worth many billions a year, is ripe for digital disruption, and offers one of the stickiest subscription revenue streams available.
2. It solves the multibillion-dollar problems of how to deliver packages with more accountability and reduce theft.

Most of us aren't home during the day, so packages are often left on welcome mats and behind potted plants, which leaves them vulnerable to thieves. If this happens (and you ask nicely), Amazon will often send you a replacement—at *their* cost. The problem is, no one knows what really happened to the original package. Did your protein powder get dropped off and then nicked by a neighbor? Did the box just get delivered to the wrong address? Or did the delivery driver themself need an energy boost from your goji berries and steal them? If Amazon could record what was happening at the door, both parties would have a better idea of who was at fault, solving a costly problem for the world's largest retailer. Add to the mix Amazon's Key product, a smart-lock system that allows Amazon delivery folks to open your door and leave the package inside, and package delivery gets even safer.

Other home security companies, like Nest, Arlo, and Google Home, are also making versions of front-door cameras that identify human movement and learn faces of people expected at your house to distinguish them from strangers. Perfecting this will benefit not only the postman, but you, too. You'll be able to set up "keys" so the dog walker can access your home, but only between 10 AM and 3 PM on weekdays—and if they take more than Gromit, you'll know. Cleaners and babysitters will have different rules programmed into the lock and be paid automatically by the number of hours they're in the apartment. And firefighters can get in anytime they want!

There is another way that Ring is different from traditional security systems. Most home setups call the police or other security services when they sense an intruder. But instead of what's happening outside your front door being streamed into the office of some bored corporate official in another state, your neighbors can review the footage through the Ring's Neighborhoods app. If the algorithm thinks someone looks "suspicious" or just doesn't belong, everyone else in your neighborhood can review the suspect and flag the behavior for others to notice and comment on. It's the 2.0 version of Neighborhood Watch.

Of course, unlike a simple tumbler lock, a security-camera algorithm isn't neutral. As with all SuperSight applications, home-security device programmers must prevent their own biases—and those of its users—from granting corporations, governments, and law enforcement undue power.

⚠ HAZARD: STATE OF SURVEILLANCE

A street camera in Amsterdam counts everything that passes by to help city planners understand the flow of pedestrians and bicycles. But did these people consent to such surveillance? And who limits its use?

Let's say you wanted to build a surveillance state. First, install wiretaps in everyone's home to listen to conversations. Then, mount cameras with built-in facial-recognition capabilities on every doorway, and stoke enough fear to inspire a volunteer network of neighbors to monitor these images as sidewalk informants. Design a system to monitor what people read and purchase. Host cloud servers and compute cycles to comb this data for correlations and patterns.

Sound familiar? Alexa is in our homes listening. Ring is at our door watching. Amazon Cloud offers face recognition services called Rekognition. Kindle knows what you read, Amazon Prime knows what you buy, and Amazon Web Services offers predictive big-data cloud analytics to determine what you might do next. Turns out *1984* was founded in 1994.

If the government forced us to place these devices in our homes, we would be outraged. But somehow we aren't as alarmed about giving up our information to a private company as consumers—in fact, we pay them for the honor.

The voluntary surveillance that started with Google ad targeting and Amazon recommender services for books, movies, and more will be small potatoes compared to the data that wearables will cull about your activities, interactions, and interests. So it's critical that we legally block companies, or divisions of companies, from connecting all these disparate data streams. This will be difficult, especially since many of these data streams are open, like your Venmo payments, and social media posts (which can include location and other information about daily activity). Marketing platforms already combine these particular obsessions and neuroses, like ice cream or eco-friendly cleaning products, and select those times when you are most vulnerable to specific messages, like after a bad night's sleep or just after a breakup, to promote their products.

A secondary effect of surveillance: computer vision camouflage makeup.

One way of mitigating these privacy concerns is to force companies to address them. Andrew Ferguson is a professor of law at American University and the author of the book *The Rise of Big Data Policing: Surveillance, Race, and the Future of Law Enforcement*. He believes we should be legislating companies' approach to privacy, and that addressing these privacy concerns is in Amazon's best interest. Amazon failed to think through many of its choices with Ring. It didn't include rules for how long they could use the footage—or how police could share it—or how to manage misuse or terms of service violations. All of these risks were eminently foreseeable and could have been articulated and addressed ahead of time. But like most tech companies in this world of big-data policing, Amazon didn't take the affirmative front-end accountability steps that any company should before rolling out a product. Proactively addressing these privacy and data-sharing concerns has become such a touchstone issue that it's now a competitive advantage to do this with transparency and strong encryption, as Apple has for unlocking your phone with your fingerprint or face.

We can also address privacy concerns by taking matters into our own hands as individuals. If you're concerned about pervasive facial recognition systems, consider adopting the anti-surveillance fashion trend set to sweep cities worldwide: weird hair and makeup designed both to defeat computer vision and get the attention of others. Brooklyn artist Adam Harvey created the computer vision program CV Dazzle, which deconstructs facial continuity through cosmetic camouflage. It's a kind of "anti-face." Similar computer-confusing methods were used during the Hong Kong protests in 2019, where people beamed colored lasers directly into the lenses of the city's security cameras to help thwart facial-recognition technology.

This is how all new technologies work: for every well-intentioned advance, there will be a nefarious actor who attempts to exploit it, and another technology that leapfrogs to patch the vulnerability and counter the effects of the first. It's an algorithmic arms race. Still, it's important to try to anticipate and forestall the hazards.

"Identifying names of other people (and pets) are what we most wish for in an augmented future."

—Pádraig Hughes, Facebook Reality Labs

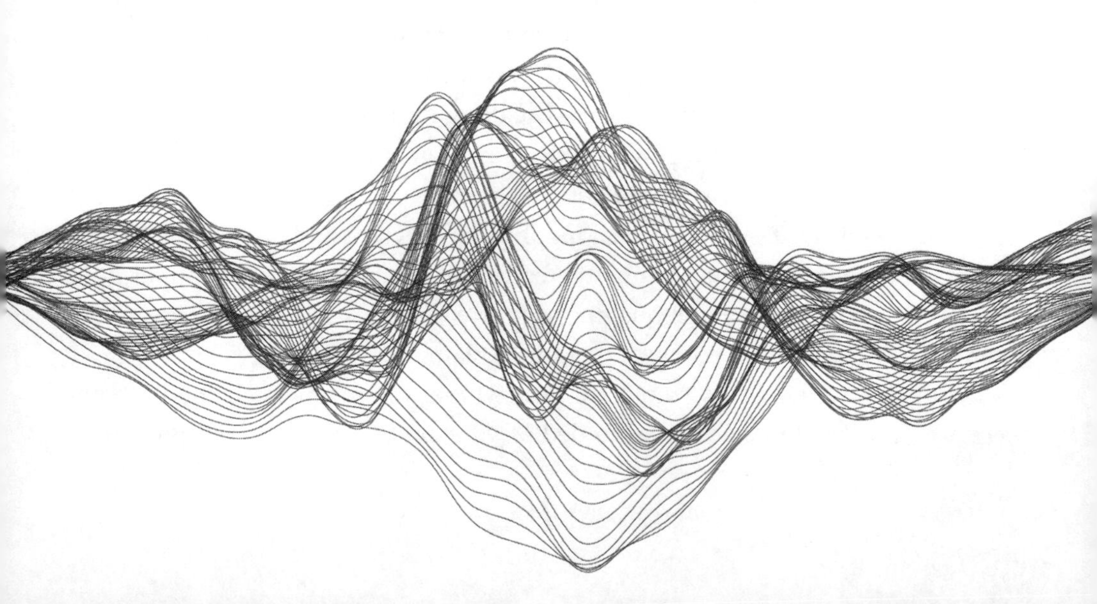

1.5 Opening doors—and your wallet—with your face

biometrics for security and micropayments

Did the doors on the Starship *Enterprise* have simple proximity sensors, or did they open because they recognized the faces of Kirk and Spock? (I'm unlikely to stop making '70s sci-fi references, so you might as well settle in.)

Computer vision enables new levels of security and automation not just in the home, but also in the whole built environment. Still-new technology that recognizes your face and unlocks your front door or phone today will be the standard lock of tomorrow. Instead of a metal key determining access, however, an invisible key is linked to your unique combination of facial features. If you have an evil twin, fear not: facial biometrics can tell identical twins apart, even if their mother can't.

Biometrics such as the topology of your face or the unique pattern of your iris are the least "spoofable" and most convenient ways to grant access. Our faces are just as unique as our thumbprints—and also a lot harder to fake. It's why your phone uses your 3D face mesh to unlock, and why airlines like Northwest are replacing boarding passes with face scans.

It won't just be airports we'll get through faster. Computer vision–enabled access means you'll soar through your whole day faster. Every hotel room, retail store, workplace, car, and bar will have a SuperSight bouncer that only lets in the people it's told are on the guest list. Think about all of the time you spend fumbling for keys to unlock your apartment, car, bike lock—everything. You won't need a flaky magnetic card to get into your hotel room anymore (and management will benefit from knowing how large your afterparty is getting). And your kids won't have to worry about anyone breaking into their diaries.

Ultimately, security is a gradient function. You trade speed and convenience against the power to identify and defeat "attack vectors" (note: I did not invent this language). That's why we have longphrasepasswords and public/private key, two-factor, pain-in-the-ass text-me-the-code-already authentication. SuperSight is the secure way out of this quagmire. It streamlines security and powers the coming personalization economy. Let me explain.

Imagine walking into your local pub where, when you catch the bartender's eye, they're immediately reminded of your preferred drink, and pour your favorite bourbon Manhattan as you settle onto the barstool—a toast to you (and a better tip for them). On the way home, after using SuperSight cameras to unlock the door as you approach, your Uber will adjust to your favorite settings: seat warmers on with lumbar support and little massage, "calm" acceleration and cornering profile, and AR projections of forest dapple light on the windows.

Pass through the huge sliding doors of any retail store in five years: cameras recognize you, consult your preferences, and start the show. Your glasses will dynamically display certain products, flag loyalty discounts, and spotlight recommended products that fit you, complement what you're wearing now, or even fill a gap it "sees" in your at-home wardrobe (more on this in chapter three, "Styled").

Billboards, shop displays, and other surfaces throughout the city will "see" you coming and, chameleon-like, change their "offers" to something that *you* find attractive, just as your social media feeds are customized toward whatever you and your influencers have been Googling—a reality anticipated in the 2002 movie *Minority Report*. My MIT friend Jon Undercoffler served as the film's science advisor and later served the same role on the *Iron Man* films. I gave him a hard time recently: "Those *Minority Report* billboards that scan your retina, then whisper your name, don't go far enough. The replicants should have been walking brand advertisements." With SuperSight glasses, these so-called replicants—futuristic "store associates"—will be dynamically generated AR doppelgangers *of you*: holographic helpers who won't leave your side from the moment you enter a store, pointing out products and amplifying any expression of interest on your part. If you don't want the guidance, giving them a *sharp elbow* gesture will shatter the illusion and leave you to browse on your own.

Stores won't be as important, however, because new products can be spatially inserted into any environment. Only a subtle glow or price tag will distinguish these product placements from reality. Much like you can get Spotify for free as long as ads can tag along, most people will opt for "free" smart glasses, where the hardware costs are subsidized by your future attention to hyper-targeted advertising. These new ads will be both more tolerable and insidious than current banner ads on web pages.

Samsung has always been ahead of the game when it comes to the future of retail. When I was an innovation advisor there, we coined the term *ambient commerce* to capture the idea of paying without friction—without even thinking.

Once stores can identify you when you walk in, personalization becomes pervasive, and any friction associated with checkout and credit cards disappears. Stores will debit purchases from your credit card, apply loyalty points, or arrange for financing automatically, without the need for cashiers.

SuperSight is removing what behavioral economist Dan Ariely calls "the pain of paying." Specifically, it connects identity (who you are) with intent (the objects that catch your gaze), so stores become totally friction free. In fact, this blows up the notion of a store at all: pick-up-and-go commerce means you can just take a mini smoothie from a Jamba Juice employee skating by, or grab a hat from a rack at the park on a surprisingly sunny day. Your credit card will get dinged later.

Ambient commerce ushers in new types of micropayments. Think of all the little tiny things you sample every week without purchasing: an apple at the farmers market, four mini-spoons of ice cream flavors before you "decide," or General Tso's chicken on a toothpick at the mall. In the future, computer vision will be counting these. If we don't want to pay individually for these bite-size products and services, we'll be able to pay a subscription fee instead. At a resort, for example, you might choose the all-inclusive option where you pay a larger fixed amount of money at the start of the experience so that everything feels like it's available to you free of charge—the same way your free-sample spree might have felt. But with SuperSight glasses, anything that's not included would simply be invisible. Alternatively, a micropayment option would tally each drink, towel, appetizer, and beach Zumba session.

I'm excited about future cities that more easily support a circular and shared economy where ownership isn't required, retail is diffused, and there is no need for stores to lock up merchandise at the end of the day. Use any *glowing bike* you find to get around, or any *pulsing tennis racquet* you see at the courts; any *shimmering surfboard* at the beach is fair game. Like those hotel sheets? They're yours! Payment per-use is low friction, secure, and automatic. Not only will you be able to walk straight into the ramen booth without a host checking that reserved table is really for you, you'll only have to pay for the amount of soup you slurp.

As you can see, computer vision's ability to recognize and tally will fundamentally change the structure of retail and hospitality. We'll be able to open doors not only with our eyes, but with our wallets, too.

1.6 Decorating the world with data (planes of projection)

We've largely been imagining the coming world of augmented vision as enabled through a pair of smart glasses. True, more than fifty companies as of this writing are developing smart glasses—and even contact lenses—that use optical combiners to embed displays in your line of sight. Each of these companies is innovating on how to make their technology as light and bright and compelling as possible. Each hopes to create an all-day display that you never take off, permanently placing you in a mixed reality. But the spectrum of options for projecting digital data onto the world is broader than this.

Apple, Google, Samsung, and Microsoft have invested in research and acquired companies to ensure that the phone already in your pocket and the tablet already on your side table support "hold up" AR experiences—just aim the device and a new reality is overlaid on the screen. The resolution and realism of these displays continue to improve, as do the developer software tools designed to help companies and hackers make 3D models, compelling games, and dynamic shopping experiences. Consumers, too, are becoming more

THE PLANES OF PROJECTION

Personal privatized view	Hold up	See through	Group around / shared view
Glasses	Smartphone	Windshield	City Scope
Contacts	Tablet	Mirror	Structured Light
Helmet		Window	
		Peleton	

Augmentation spans a continuum between personal and private smart glasses, and shared and collaborative data projection.

comfortable understanding and interacting with this SuperSight universe, from the Pokémon we catch while walking the dog to the face-decorating filters we apply on Instagram and Snap. And the biggest step in adoption just launched in mid-2020: Google now offers AR content in search results. Try it now: search for *shark* and press the AR icon to pop an animated shark out of your phone into the world around you!

But as big as your tablet is, some situations demand an even larger augmented window—like the HUDs in some high-end car windshields, airplane cockpits, and the Starship *Enterprise*. When split seconds count, these displays reduce the time from information to action. No need to avert your gaze and scan other instruments or screens before making a decision. You may have experienced an augmented overlay in the form of the backup camera view on recent cars. As you shift into reverse, you see two layers of information on the monitor: the actual camera feed and a diagram that changes as you turn the wheel. The combination shows your turn radius, so you don't clip another car or a nearby post in the parking garage.

Over the next few years we'll start to see augmented windows in skyscraper viewing decks and high-end hotel suites as a flagship feature to complement the panoramic view. Look out across the city and see annotations with the names of buildings, parks, and routes to walk, as well as sponsored destinations like restaurants and theaters. Instead of crowding around a single phone with your family, trying to work out the best way to get to Bryant Park and where to eat nearby, you'll gather around the window to plan your route, and swipe to browse which restaurants have the best gluten-free specials.

Next will be department-store windows. Rather than walking by stationary (or animatronic) mannequins wearing outfits, you'll see coats-of-many-colors, glasses, hats, jewelry, and more immediately superimposed on you while strolling by.

With augmented windows in cars, buses, and trains, there will be opportunities to learn more about the scenery passing by. Imagine a tour bus with data projection on every window that calls attention to monuments or restaurants as you drive around a new city; you might wink or make a quick snapping gesture to bookmark those places to return later. While it would require a very bright projector to overcome the bright scene, a very large heads-up-display is convenient because anyone on the bus can use them without special glasses, and they could sync to your Google or Tripadvisor account when you tap your phone.

SuperSight windows create a mixed-reality experience that is more compelling because it is shareable—others are able to see the same augmented reality you do. But windows have parallax issues; the augmented layer needs to be positioned to align and register with your vantage point, the line between your eyes and your view of the world. Ideally, objects would be painted directly with digital information, so anyone standing around them sees the same overlays mapped directly on the object.

Welcome to the world of data projection. When I was a student at the Media Lab in the 1990s, my favorite demo (which is saying something, because it was a place with many, many demos) changed how I thought about augmenting ordinary objects with information forever: a light bulb with SuperSight.

You approach an all-white drafting desk with a single, white light bulb at head height, pointing down to the surface. Nothing out of the ordinary here. But when you place a small architectural model of a house on the desk, something incredible happens: a long shadow animates out from the house by four or five inches, like a shadow cast by early-morning sunlight. Next, you grab what looks like a little clock and also place it on the surface of the desk. As you turn the clock to a different time of day, you see the shadow around the house change direction and length; it's longer near dusk and shorter in the middle of the day. Now you grab another tool that looks like a little arrow vector. When you place that under the bulb, little strands of wavy lines are projected around the house, showing the direction of the wind.

This is fun, so you decide to add another building, and suddenly the simulation changes. You can now see how a wind tunnel effect is created when two buildings form a narrow gap. (In my home of Boston, the unconventional parallelogram shape of the Hancock Tower downtown creates a dangerous wind vortex that nearly knocks down people crossing Copley Plaza.) Playing with the two structures, you also see how the shadows from one building interact with the other, becoming shaded when they are too closely spaced. You are interacting with a dynamic simulation that is reading the positions of the buildings, the time of day shown on the clock, and the direction of the wind vector arrows in a way that city planners only ever dreamed of—until now. It feels like a magical sandbox.

This IO bulb (the IO stands for *input and output*) has an ordinary Edison screw socket, but it's not an ordinary light bulb. It's creating what's called a "dynamic texture map" on the desk's surface, rendering effects on the house model below it. The sensor for the IO bulb was a little camera placed just adjacent to the projector, able to read the position of architectural models, the time on the clock, and your hands in the scene. Using computer vision, it calculated the positions of all these as inputs and generated the appropriate output: a shadow, wind vector lines, and building texture.

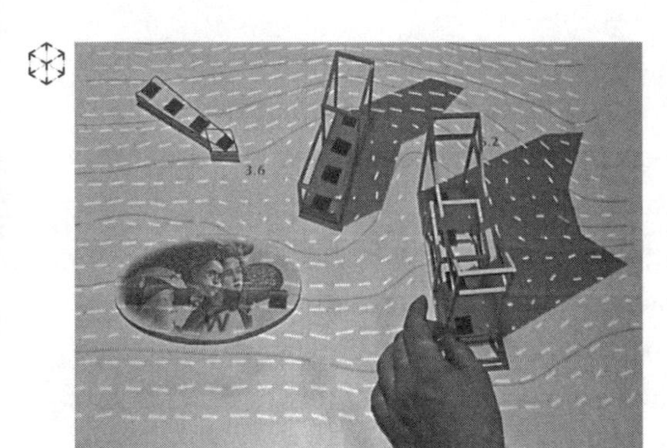

The IO bulb is a conceptual extrapolation of the common lightbulb in which the incandescent filament has been replaced with a camera and projector. The bulb "looks out" into a region of architectural space, using the camera to analyze activity within the space and the projector to emit visual information back into the space.

While I was an instructor at the Media Lab, I saw the simple IO bulb concept develop into a city planning product to envision the future of cities. It was used to plan emergency refugee housing in Germany, to craft a disaster plan for the tiny country of Andorra, to encourage mixed-use amenity zoning for Cambridge, Massachusetts, and more.

For example, to study the effects of ride sharing and autonomy, the Media Lab's City Science group built a data-illuminated physical model of Kendall Square near MIT called CityScope with the fastest building material around: LEGO bricks. Each element of the miniature city model was machine readable, so a computer could construct an accurate replica as people reached in and moved things around. The digital version recalculated metrics important to city planners each time someone moved a park or increased the height of a residential tower, then rendered those metrics over the physical model as a series of animations and heat map. This immediately displayed the results of a policy change or physical reconfiguration of the space. Trading parking space for housing density, for instance, produced neighborhood vitality, restaurants, and nightlife. Anyone gathered around could pick up and move buildings, create bike lanes, or remove parking structures, and then see the effect on the future city's walkability scores, diversity, density, and sustainability.

This projected-light augmentation is extraordinary because no glasses are required, the experience is shared, tangible input and immediate output are strongly coupled, and results are represented directly in context. A group of people gather around a model, move elements of a city around, and see the resulting cascade of consequences for this action.

Now imagine if every old screw-mount Edison light bulb in your office and home were replaced by a bulb-sized data projector that could display a 4K-resolution movie on any surface, dynamically skewed and warped to accommodate that surface's geometry. With a mesh of these overlapping data projectors, and a model of the environment, every surface and object would become paintable with pixels: hospital and hotel floors with navigation cues, conference room and office doors with schedules, restaurant tables and bars with reorder buttons, and coat hooks in your home with jacket suggestions. And yes, book spines in your local bookstore will show animated author profiles. In a projection-rich environment, product packaging doesn't need print, because it becomes a dynamic canvas to wrap with personalized video.

Projected light is typically location specific; light bulbs don't walk around. But projectors are becoming more portable, even wearable in a necklace or hat. This opens new possibilities for painting the world with information.

A master's student at the Media Lab, Pranav Mistry, prototyped a system like this called SixthSense. He basically wore an IO bulb as you would a name tag. As he walked around the world, the camera detected what was in front of him, and a pico-projector illuminated objects in front of him with information. As he lifted up his hand, a watch was projected onto his wrist with updated airline boarding gate and times. He demoed a shopping experience where Amazon ratings were projected onto items in a grocery store, and star ratings appeared on the toilet paper.

Similarly, a team of students from the MIT Enchanted Architecture class that I co-taught with Gilad Rosenzweig, developed a gurney that helps emergency responders quickly find their way through a complicated building. A data projector on the gurney projects large arrows onto the floor and walls to guide them

The SixthSense Projector from MIT allows a wearable projector to augment everyday objects.

to the patient. On the way back, the projector displays the patient's vital statistics along with navigational cues. The advantage of this projection over smart glasses is that all the emergency responders share it, allowing them to react in unison.

The path-projecting gurney is a larger version of one of my favorite AR projection projects, the Magic Flashlight. The assignment was to prototype a tangible tool that made something invisible visible. A team of students found the architectural schematics for the recently finished Media Lab building at MIT—CAD models that showed the plumbing, air ducts, electrical wiring, and hundreds of miles of Ethernet cable all snaking through the walls of the building. Then they packed a vintage flashlight with the encoded building information model, a pico-projector, and an IMU, or inertial measurement unit. This combines an accelerometer, gyroscope, and magnetometer/compass—components that determine where you are and where you are pointing, and are standard on phones and smart glasses. The interaction was amazing: flicking on the Magic Flashlight revealed what was inside that wall. You expect light when you turn on a flashlight; you don't expect X-ray vision.

I hope your mind is spinning now, thinking about how the ground plane of the ordinary world will soon get another layer of information projected directly but selectively on top of it, positioned and skewed in response to the surface beneath, oriented and styled in response to you. SuperSight's ability to see and recognize what is in front of us, then direct the spotlight of our attention, gives us a new way to interact with the world and each other. We can tell corny jokes about guitars, let the dog walker into our house automatically, see ratings while shopping, and even see inside of walls. The source of this augmented layer can be private (in your glasses or through your phone), shared (on a fixed surface like a window or windshield), or projected directly on the objects in our environment to change how we understand the world and make everyday decisions about, well, anything.

We've looked at what SuperSight will let you see in the future—but what about how *you* will be seen in this new, more revealing light?

What if a digital personal
assistant could see what you see?

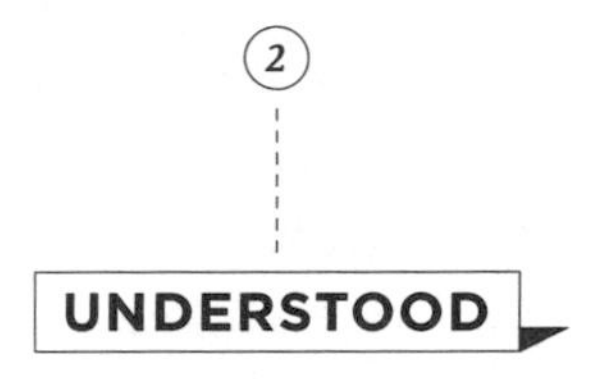

About living with a set of coaches who analyze, guide, and advise you

2.0 Your personal guardian angel

five types of coaches we all need

The idea of personal angels was first introduced in *The Shepherd of Hermas,* an early Christian book written around AD 150: "There are two angels with a man—one of righteousness, and the other of iniquity." Hermas is told to listen to both angels but to only trust the Angel of Righteousness (good advice).

As it turns out, nearly every early society had their version of this kind of angel. Personal "tutelary" spirits, whose role is to keep you safe and guide you in the right direction, are common across ancient and traditional cultures. In late Roman times, for example, they had the *genius,* a personal deity that followed one from birth to death; the Greeks had *daimones,* lesser deities or guiding spirits (and the inspiration for Philip Pullman's animal daemons). In Christian folklore, each person has a dedicated guardian angel to prevent them from coming to harm—both physical and moral—because we are being constantly assailed by tempting devils. Japanese Buddhism has a pair of beings called *Kushoujin* that sit on a person's shoulders, recording deeds both good and bad. According to Islamic literature, every person has two *qareen* (literally "constant companions"), one from the angels and another from the "jinn"—which eventually became Anglicized as genie, the wish-giving supernatural being of Persian folk tales.

A guardian angel may be the perfect metaphor for a personal wearable AI coach. The smartphone assistants we now keep in our pockets do some of the work of angels: they log our deeds (both naughty and nice), and introduce ideas

Baroque painting of *The Guardian Angel* by Pietro da Cortona, 1656, from the National Gallery in Rome.

(or at least fact-check our friends). With SuperSight, they will become more knowledgeable, intimate, and influential: beyond counting our steps, logging our whereabouts, eavesdropping on our conversations, tallying what we buy, and monitoring our cough, we will grant them the gift of *sight*.

If you had an omniscient angel sitting on the temple of your glasses, what wishes might this sprite grant? For safety? For morality? For vigilance or temperance?

Before a new guardian angel can offer advice, it must first understand through observation. So, learning our daily habits—our eating and exercise patterns, our media consumption, our sleep hygiene—is its first job. Camera-enabled supervision services will see the world as we see it *and observe us* as we interact with others. As designers, we then need to craft *how* it should intervene and influence, if at all. Should the tone and presentation of this feedback feel assertive and critical like a drill sergeant, affirming and ethereal like a new-age yogi, or Socratic and insightful like your best professor? I hope it's like a supportive coach. These new genies will understand us—and, in turn, help us understand ourselves.

Why do I believe that our relationship with technology should evolve to feel like one with a coach? Like most coaches, technology knows more than us and has seen more than us, but should ultimately let us play our lives, win or lose. It isn't either's job to puppet us, but instead, to have an uncanny ability to whisper the right words of encouragement or caution at the precise moment we need them. Sometimes they motivate us; other times they hold us accountable for our mistakes. We rely on them to keep us from harm or nurse us back from injury. While their focus is on how we move through the world and interact right now, they also see something greater for us: a vision for our more noble/performative/empathetic/creative/healthier future self.

Rather than having one meta coach, we'll amass a community of coaches, each with a different voice, personality, and access to certain types of data, for which we'll have various expectations about the role they play in our lives. Here I'll introduce you to five (though we'll touch on just a few in more detail, later in this chapter):

- **Performative coaches** help you improve how you move, whether playing a sport, exercising, or dancing. These coaches rely heavily on something called pose detection, which sees where your body is in space, how fast you are throwing a ball, and if you need to focus more on your pivot for your right jab.

- **Wellbeing coaches** help you track your sleep, hygiene, stress management, eating habits, and meditation practice. These coaches examine your facial expressions to determine your mood and stress levels, and use heat cameras to infer sleep stages or illness, activity classification to understand lifestyle choices, and food recognition to log what you eat.

- **Organizational coaches** help you break down goals into tasks, then delegate, organize information, plan, and schedule time with discipline. These coaches help you focus on a task, regulate your attention, and reduce the cognitive load of a world full of often meaningless choices.

- **Interpersonal coaches** help you learn leadership and social dynamics, emotional empathy, and conversational style. These coaches study your listening skills, how you affect and influence others with your voice, tone, and posture—and can help you keep that first date around longer.

- **Existential coaches** help you ask important questions about your values, your priorities, and your purpose in the world. They encourage you to prioritize the right things and understand the long-term consequences of your actions, while keeping your mortality in perspective.

While these types of coaches are likely familiar, the technology of SuperSight will make each more helpful and accessible to a broad audience. They'll also become more insightful the longer you employ them, because the more they observe your patterns, the more they know your strengths and gaps. And the more perspectives they have on the action, the better. You'll thus want to give them access to the cameras in your life: the fixed ones in your home and work, those in your car, the ones you wear in your glasses, and even those embedded in your friends' and coworkers' glasses.

Once they process both the scene in front of you and how you appear to be reacting to it, they will deliver their feedback judiciously and subtly, whispering advice in your ear: "Your boss is confused—maybe try a different example," "Don't forget the paprika for that goulash," "Try twisting your left hip a little more to get another fifteen yards on that golf drive," "Shut up about yourself and just ask another question about her."

The best coaches will have a personality and use metaphors to inspire change. Instead of nagging comments like, "You aren't very flexible in this yoga session" or "You're singing flat again," they'll offer mantras or metaphors: "Your body is a flowing river" or "Imagine the tone spinning up from the top of your head." Delighted, you'll think, "Oh, I *can* do that!" Our SuperSight coaches will also evolve to use humor and novelty, too. If they don't, we'll banish them back into the bottles we rubbed to summon them.

2.1 Understanding your form

3D pose detection *gamification* *moneyball scouting*

It's not easy in basketball to make a three-point shot; even the best high school players struggle with it. Experienced coaches help by analyzing their players' shots and offering advice: "Make sure your elbow is above your head when you release the ball," or "Try to release the ball at the full extension of your arm." These adults have seen many more misses than their teenage players have, and their advice can be the difference between a college scholarship and mediocrity.

Soon, however, you may not need a human coach at all. You can hire a SuperSight coach to improve your shot instead.

SuperSight unlocks a new generation of computer vision–enabled autonomous coaching services to analyze how you currently perform and help you level up to your aspirations. Using the built-in cameras in your mobile phone or computer, an AI can study a slow-motion video of your technique and compare it to your desired superstar player. Want to be like Mike? The system can line up your free throws against the mighty Jordan so you can see the right angles you should hit with your elbows.

Just how big is the market for sports coaching? I did a little entrepreneurial market sizing for you: worldwide, 65 million people play basketball, 75 million play tennis, and 220 million play badminton (surprise!). Yoga has 300 million practitioners and is growing gangbusters; Ping-Pong also has 300 million players, plus volleyball has 800 million (really!). And leading the participatory-sport pack? Soccer, with 1.3 billion players worldwide. Add to that the fact that, thanks to the COVID-19 crisis, everyone who typically went to the gym or did group exercise classes had to adapt to working out at home—often at a fraction of the cost. How many of us will go back to stinky locker rooms when we can get top-notch coaching in our living rooms?

My favorite computer vision coach, HomeCourt.ai, uses numerous gamification strategies to motivate you and your child/partner/roommate to jump as high as possible, shuffle back and forth to tag virtual cones, dribble a soccer ball while hitting virtual targets with your hands, or dribble a basketball while doing math or word games. It features celebrity basketball coaching videos, and uses computer vision to count your three-point shots or HIIT reps. It requires no special hardware—just lean your phone or tablet up against a wall—and most activities are free. What's not to like?

HomeCourt tracks my son's soccer ball to improve his agility and ball-handling skills.

In 2020, as a response to the sedentary working@home inertia, I created an AI coaching app for yoga and other sports at Continuum. As a futurist, it's my job to imagine projects at the middle of a Venn diagram of trending customer desires, cutting-edge tech, and business viability. With such a huge portion of the world normalized to working out from home and such powerful computers in our pockets, this felt like the perfect opportunity to have some fun.

POSE DETECTION

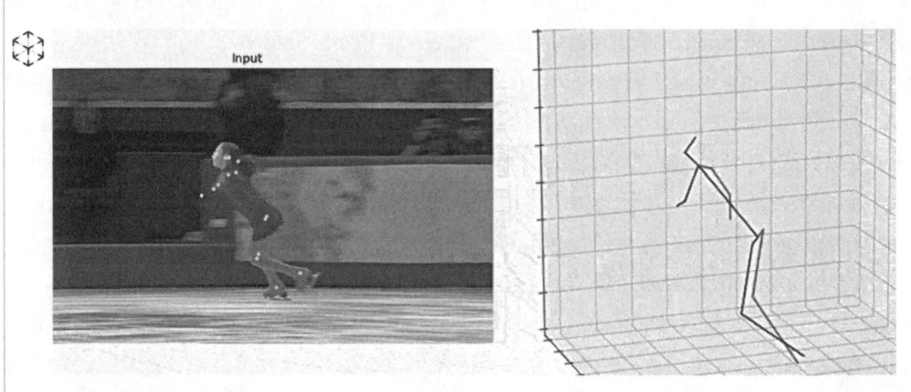

A 3D pose classifier is a type of neural network that infers the position of the body in motion.

The key computer vision technique that SuperSight sporting coaches use is called "pose detection." A neural network is trained to identify each part of a body, and determine the angle of the thirteen primary joints (neck, shoulder, elbows, wrists, hips, knees, ankles) and even the position of each finger. Ten years ago, this required an infrared projector and camera system, which had a range limited to a few meters—think Xbox Kinect sitting atop your television screen while you boogie in your living room to *Dance Dance Revolution*. Now, specially trained neural networks use ordinary camera optics to detect the human body in 3D motion with increasing precision and at ever-farther distances.

Using normal cameras feeding footage into computer models, we can now detect twenty soccer players on a field at 120 frames a second, even with some players overlapping other players. We can discern not only where they are on the field, but also the angles of the key joints of their bodies, which direction they are looking, and the power they are exerting. As this technology gets better and cheaper, access to these systems is quickly migrating from the pros to college teams, and will soon be coming to tennis and golf courses and backyards, where we'll benefit from the same embodied feedback as the best athletes in the world.

At the time, we were working on using computer vision to analyze plays and performance for a professional sports team, which sparked the idea of creating a more general tool. On this project, our challenge was to take video of flow-based games like soccer, then track players and label specific moves—dribble, pass, shot on goal—to identify opportunities for coaching or spotting talent, *Moneyball* style. And as we observed and interviewed players and coaches, we discovered a key insight: comparisons are the most efficient way to learn and level up sports skills.

Nearly every design discipline pins up side-by-side comparisons. Look around Continuum's Boston studio and you'll see multiple options for branding, landing pages, and app flows covering the walls. The design-thinking process requires generating alternatives and carefully considering them side by side to weigh differences, discuss trade-offs, and find ideas for improvement. The best sports coaches and camps use this same technique with videos of all-star players to find gaps in technique and form. When players observe these differences, some ideas for improvements might be obvious to them, but coaches help draw attention to the most salient next steps. Might SuperSight be able to focus players on opportunities for improvement in this same way?

Not everyone can afford a professional coach for their at-home workout routine, though. How might we give people exercising remotely the type of form and technique guidance you might get in a yoga class or with a personal trainer?

Continuum's prototype yoga coaching app with pose detection.

There are a million yoga videos online and more being posted every day by Equinox, Orange Theory, and gyms on their Instagram channels . . . but none of them offer personalized feedback, because they can't see you.

The yoga coaching app we created experiments with this. First, you pick an expert and an action—for example, Baron Baptiste's triangle pose—set your phone down, and record yourself doing the activity. The app then uploads the video to the cloud, runs a pose-estimation neural network, aligns the motion, fills in more frames to generate super-slow-mo, and voilà—you can see where your pose matches and differs from Baptiste's, and how to improve.

Quantified workouts are a rich and exciting field for entrepreneurs across every sport. But there is definitely such a thing as too much information. For example, back in early 2020 I spoke at a healthcare conference in Jackson Hole, Wyoming. While there, I spent most of my free time on the slopes. A European company had sent me a set of "cookies" to stick to my skis that would measure about a hundred variables and provide coaching advice on my carve radius, body angle, ski symmetry, turns per minute, fitness, and more. Sounds great, right?

I was surprised by how much I didn't like it. The data was so complete and detailed that I found it both overwhelming and unwelcome. As a general personality style, I invite feedback and constructive criticism, but this was like a speech where you were scored on each individual word, rather than the meaning or intent behind what you were saying. I did learn a few things—I often lean back when I'm descending a steep slope (a classic anxiety mistake . . . that I apparently made 378 times one morning)—but overall, the data felt like micromanagement from a boss that should have had better things to do with their time.

Whether you're coaching your kid's hockey team or commenting on a friend's volleyball play, how much feedback is optimal? And do you call attention to what is done right, or correct mistakes? A critical approach where you list all the problems might feel more efficient but overwhelms the user with data, monopolizes their attention pool, and is often worse than no coaching at all. When self-confidence is at stake (and when isn't it?), the best coaches emphasize positive behaviors first and simplify what you should work on next, with statements like, "Everything looks great except I want you to think about this one thing . . ."

I once went to a workshop led by legendary tennis coach Tim Gallwey, who wrote *The Inner Game of Tennis*. The metaphor he used to describe effective coaching was *a spotlight of attention*, or focusing attention on a single aspect of performance, rather than instruction: "Think about where your racquet is as the return comes over the net" rather than "Get your racquet back earlier."

You often know intuitively what "right" feels and looks like—the topspin backhand cross-court shot, the all-net-three-pointer. Sometimes all you need to fix your form is someone—or some SuperSight-driven AI—to direct your attention.

⚠ HAZARD: COGNITIVE CRUTCHES

For many years, athletes have tried to outdo the competition by taking performance-enhancing drugs. Could SuperSight coaching convey the same kind of unfair advantage? Coaching certainly will give individual athletes a distinct edge, but because these AI services are automated, they generally will be affordable and accessible to most.

Sports teams using AI-powered coaches need to set fair policies about data access, ownership, and persistence. Leagues must safeguard data and be clear about transferability to keep players on a level (literal) playing field. Data is increasingly used to benchmark athletic performance, predict injury recovery, and scout future all-stars. Once recruited or traded, should a player's data automatically transfer to the new club, or should data be sold separately?

We also may form cognitive crutches because of our excellent AI coaches. Calculators and GPS caused a generation to struggle with working out 15% tips and wayfinding in foreign countries without a data plan. So as not to become overly reliant on scaffolding services, we will all need periodic analog sabbaths: no-coach cleansings to recognize our own agency and dependencies.

Might we become so dependent on conversational scaffolding and prompting that our interpersonal skills atrophy?

2.2 Finessing fitness

types of coaching feedback | *bespoke AR for sports*

My passion for cycling started early, with banana seats and coaster brakes. In high school I cycle-toured in France (what were my parents thinking?) and worked in a bike shop called Yellow Jersey in Madison, Wisconsin, a Midwest biking mecca with a 2:1 bike-to-car ratio. Of course I became a superfan of the Tour de France; we had it playing nonstop on a little TV in the shop. The world's best cyclists scale the highest peaks in the Pyrenees and Alps on sleek prototype machines, then descend at over 60 mph, with inevitable pileup crashes featuring instant blood, gore, splintered carbon frames, and folded wheels. They race an epic 120 miles per day, every day in July, finishing in Paris. It's the highlight of the summer for millions of wildly passionate cycling fans. They flock to the route, forming narrow passages for the bikers as they ascend the highest peaks, and ringing cowbells as glimpses of color stream past. Rabid fans run along next to the frontrunners on the steepest climbs with team colors war-painted across bare chests and their horned Viking hats flying.

Well, thanks to SuperSight, that cycling rapture has been virtualized, distributed, and made participatory for you and me. Now I have my own crowded racecourse in my home, populated with real cyclists to draft behind and pass. Last week, 3,804 bikers circled a virtual course through the Alps—all over the world, at 7 AM EST on a Wednesday! Each participant pedals their race bike on rollers or clamped into a trainer for variable resistance and tilt based on the terrain, and

Thousands of bikers have motivated each other to get real exercise in the fantasy world of Zwift.

even a variable fan to represent your speed. The visual experience projects on a TV, or ideally an immersive headset. This massive race is powered by Zwift, which charges $10 a month for unlimited sweating. You earn better virtual gear and more colorful jerseys as you log more miles. I admire the creativity of routes to ride: there are urban and rural loops, of course, but there are also glass tubes forty feet wide that snake under the ocean, and fantastical skyway trails that loft hundreds of feet above the city supported by . . . who cares, it's a virtual landscape!

The point of this massively multiplayer online sweat is motivation. Immersed in Zwift, the puddle of sweat on the floor and lactic acid burning in my legs is beside the point; I'm keeping up with some rider from Italy wearing a yellow jersey. I ride longer and more regularly, and I'm climbing higher virtual summits than ever.

In 2013, another cycling startup launched a home exercise bike on the crowdfunding platform Kickstarter. By transplanting the energy and experience of a spin class and the accountability of a coach into people's homes, it started a sensation in home exercise and launched a billion-dollar revolution in remote workout formats. Central to Peloton's playbook is the scheduled group workout: as you sit on your bike, a large screen teleports you from your bedroom to a live studio event in New York or LA, where real and beautiful people are sweating, struggling, and rocking out to Beyoncé. Soon, two crucial social dynamics will make this experience even more motivating and immersive: SuperSight glasses will spatially situate you between other riders in the class (you'll see them on your left and right), and your sweaty spin-mates will see you.

Inspired by Peloton's spinning success, in 2018 an entrepreneurial rower in Boston launched a Kickstarter for a product he called Hydrow to augment the traditional rowing ergometer. After coaching the US rowing team to a bronze medal at the 2015 World Championships, Hydrow's founder, Bruce Smith, was the executive director and force behind "community rowing" on Boston's Charles River, where my daughter learned to row and fell in love with the sport. We first met right after Hydrow had raised their Series A round investment. Like Peloton, Hydrow had designed a high-quality piece of "exercise furniture" with a large screen to bring you into the action. But instead of the neon-tinted club-like environment that has become synonymous with spinning, rowers scull down the Thames past Big Ben, or through the mist of a crisp, fall-foliage morning on New Hampshire's Lake Winnipesaukee to views of the White Mountains (my fave). Hydrow also employs a dozen Olympic rowers to host the workouts, and they are diverse, entertaining, and ridiculously telegenic.

Like Peloton and Zwift, the future of Hydrow will be more collaborative and connected—immersive, personal, and motivating. Connected spatial audio between people in your boat will let you hear each other breathe. Breathing together while exercising turns out to be remarkably motivating and pleasurable; as anyone who has played a wind instrument in an ensemble (for me it was trombone in a high school brass quartet) or sung in a choir (I'm thinking of the synchronized breath before the "O Fortuna" section of the *Carmina Burana*) knows, breathing together can be exhilarating, even spiritual. Also, a good rowing crew benefits from being precisely synchronized on the catch (when the oars enter) and finish (when they exit), and breathing together helps coordinate this action. Soon, projected-light AR will flood your living room with the sparkling water of the Charles River as bridges fly by overhead, filled with race spectators and spatialized audio cheers, and other boats will overtake you when your pace lags. You and your team will be motivated to take the stroke rate higher together because it will feel like you're participating in the "Tour de France of rowing": the Head of the Charles Regatta. Power ten!

I'm fascinated by the pro-health potential for these tele-workouts, which mix real and virtual coaches, teammates, and courses. The Hydrow team is doing what every healthcare company should do: offering a coaching service that feels personal, authentic, and so persuasive that you will engage day after day. Sure, it uses a physical rowing machine to spin its immersive magic. Still, the more enticing aspects of the service are the relationships with Olympic athletes, live

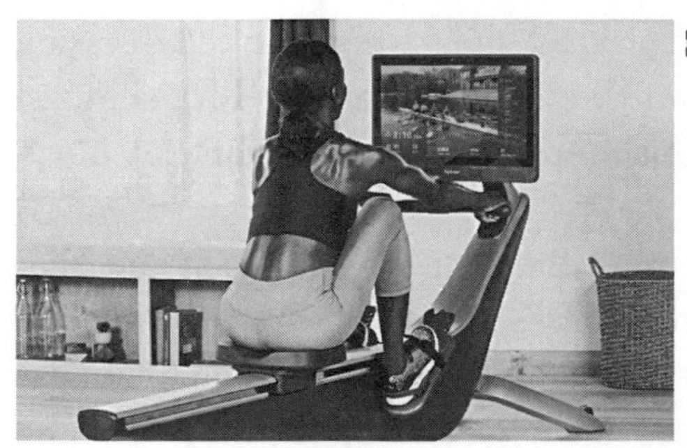

Hydrow provides a live-streamed rowing experience: just synchronize your stroke rate and you feel like you're in the boat.

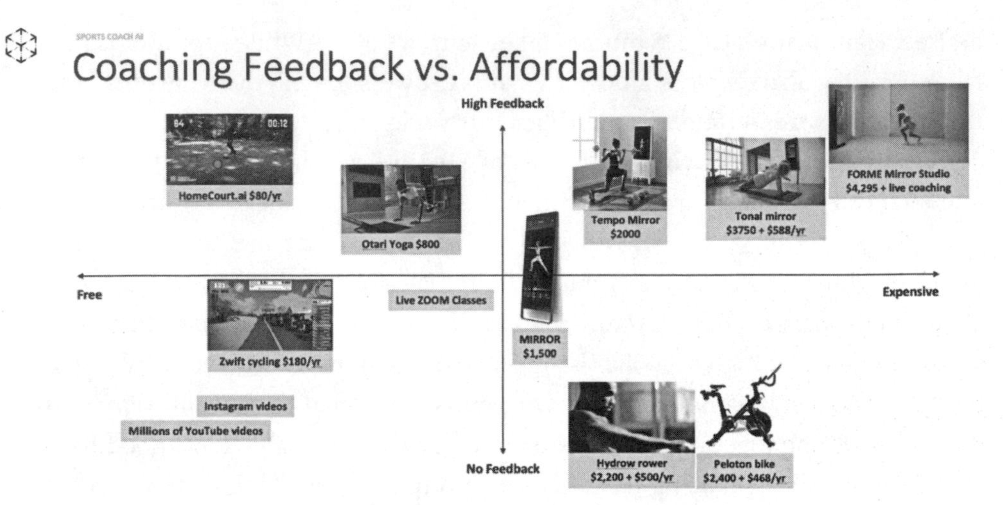

There is a large opportunity for sports brands and star athletes to develop accessible coaching services with personal feedback using SuperSight.

aspirational rows down rivers in Europe, competitive leaderboards, streaks, and rewards for rowing your first 100,000 meters—all for only $38 a month (though the machine costs $999).

If a rowing machine is too big for your one-bedroom apartment, there are now four well-funded companies hoping to embed an AI coach into a large mirror hanging on your wall: Mirror, FORME, Tempo, and Tonal. All use computer vision to provide feedback for your moves: yoga, weight training, HIIT, or Pilates, respectively. You summon your virtual trainer with a tap on the screen or an app to start a workout. They introduce moves, count your reps, offer supportive feedback, and recommend adjustments if you're not leaning far enough back for squats or keeping your back straight on kettlebell swings. As FORME founder Trent Ward told me, they believe their product should replicate a 1:1 session with your personal trainer at the gym, with workouts that are newly customized each time, and feedback that feels personal, sophisticated, and emotional, much like a trainer's voice.

FORME's particular "decentralization of the gym" is high touch and high cost. Trainers—the live human instantiations—need a video studio in their home. You pay about the same $75 personal training session fees you might pay at a gym, and the company takes a cut off the top. At that price point and level of personalization, I asked Trent if the physical therapy market might be a better target, since insurance covers these sessions and it's challenging to get into a car or bus to get your body to a clinic if you have a new hip or shoulder. "Yes," said Trent. "That's in the roadmap!"

Mirror, now a part of Lululemon, superimposes your reflection on a recorded coach and uses computer vision to count reps and evaluate your technique.

Mirrors are an excellent, compact, home-friendly interface for fitness, but if you're out in the world cycling or swimming or snowboarding, you need a wearable coach. Today, a few companies are developing niche hardware + software solutions for specific contexts. In many ways, AR motorcycle helmets, construction helmets, and ski goggles are easier design problems than creating all-purpose, wear-all-day AR glasses—the requirements are more straightforward, and the marketing channels and messaging more focused.

RideOn, ironically based in sunny Tel Aviv, where there is no snow but many aggressive tech entrepreneurs, has developed SuperSight goggles for skiers and snowboarders. They feature tracking for brag-worthy snow stats like carving radius, total vertical feet, and max air-time for jumps, an HD camera for video recording, and a resort map. And since skiing is so social, there's a find-my-friends locator and a game where you slalom through virtual gates down the real mountain and compete with others. But the killer app I envision is an alpine Waze to predict which gondola has the shortest lines or guide you to the slope-side warming hut with the best apple tarts and cheese fondue.

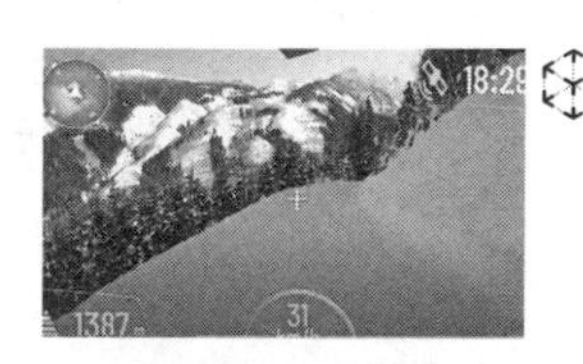

Snowboarding with speed, elevation, and air-time data from RideOn's AR helmet.

Compare these AR ski goggle features to another wearable piece of SuperSight hardware, the AR FORM swimming goggles. Both are waterproof and use a HUD projection, but their users' needs, and therefore their features, are very different. People don't get lost while swimming laps, or stop for fondue, so FORM gear offers no

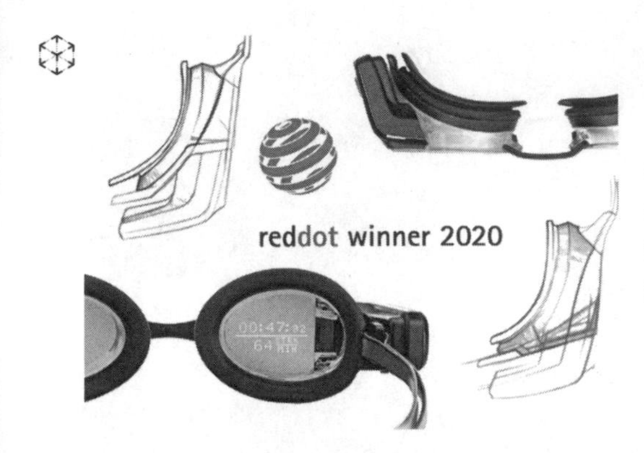

FORM SWIM goggles with AR feedback on stroke efficiency and heart rate. Swimming is typically visually dull, so I'm glad to have the feedback and distraction.

map. And swimmers care about their pulse, stroke rate, and distance per stroke, not altitude and avalanche risk.

Of course, hardware will continue to improve, but the larger, durable business opportunity is the personal coaching experiences that gear can deliver. Seasoned sports stars will brand and launch coaching apps that download into glasses and goggles, permitting a loyal fan base of aspiring athletes to learn insights, receive advice, and hear stories as the celebrity athletes reminisce about their own rise to stardom (everyone likes a hero's journey). Learn to play tennis from a projected Serena Williams or try to sprint alongside Usain Bolt—sorry, you probably can't clock 27.8 mph even on your e-bike. Instead of a life of breakfast cereal and sneaker promotions, these stars will have a more meaningful and durable legacy. Celebrity athletes will live on as personal projections for millions of young aspiring athletes, who will select their ideal mentor. While working with them over months or years to hone their skill and mindset, the young athlete will learn their hero's philosophy of life and attitudes towards sport and competition—like a long-running podcast, but one empowered by SuperSight to be more coach than host.

ROTOSCOPING, GLOWING PUCKS, AND NO REFEREES REQUIRED

How augmented reality changes spectator sports

The yellow 1st & Ten line was first broadcast by Sportvision during ESPN's coverage of a Cincinnati Bengals–Baltimore Ravens game on September 27, 1998. The line is seen only by the television audience and seems to be magically painted directly on the field, under the feet of players. It makes the game more legible, exciting, and precise than the old orange signal poles, connected with a ten-yard chain, that are shuttled back and forth on the sidelines. Over time, other LVI (live-video-insertion) systems, as they are called, have helped spectators track action that's hard to see by using glowing hockey pucks and golf balls with parabolic trails, officiating the action with baseball-pitch strike-zone boxes, and even aiding the understanding of complicated rules like offsides in soccer and hockey. World Cup sailing has been transformed into a spectator sport by augmented layers that make its byzantine rules legible even across the expanse of open ocean. Now spectators can see who is in the lead and who has the right of way that will block a key tack around the last buoy.

Pose detection helps a baseball coach see the details of form, and make a better decision about when to swap in a replacement.

World Cup sailing is now a spectator sport, thanks to rotoscoped explanatory information.

Recently, SuperSight has been adding even more augmented layers of statistics and letting the audience decide how much they want to see. During Tour de France bike races, we can now see biometrics of some athletes ("The heartbeat of the leader at the head of the Peloton just spiked to 170 BPM to maintain that position!"). And for the betting audience, augmentation can show risks and probabilities like the optimum statistical distance for a football placekicker's field goal range.

Sports broadcasts are also about to be radically personalized, based on your level of sophistication with the game or which of your fantasy teams will be affected by the next play. You'll see names of players, biometric data, velocity vectors, predictions on where they'll pass next, and even points of views on the field once thought impossible to capture: from any player, field goal post, or the ball.

Court Vision is one example of this technology. It was introduced by the NBA's Los Angeles Clippers in 2018 as "a revolutionary new way to watch Clippers games that puts fans in control of the viewing experience," by rotoscoping (superimposing) player stats, shot probabilities, ball trajectories, and lighting-effect animations for every three-pointer or dunk.

These systems take what used to be done on chalkboards and superimpose it on the field of play. With SuperSight HUDs, sports may become more dynamic, and when AI coaches show athletes real-time strategies for where and when to make their moves, this view might be shared with spectators (with the exception of baseball, where sign stealing is a taboo). The goal for sports leagues is to increase viewership, so these information-rich systems are likely to be adopted across more sports.

2.3 Seeing feelings

affective intelligence *wellbeing coaches*

The academic field of emotionally aware systems was pioneered by Roz Picard. She wrote the book *Affective Computing* and runs a group by the same name at the MIT Media Lab. For a decade, her students have developed wearable sensors that measure galvanic skin response, which is how the salinity of the sweat on your skin changes in predictable ways as you get more excited. They have recently trained computer vision systems to look at facial muscles to measure microexpressions that reveal emotional states.

One semester when I was teaching at the Media Lab, two of Roz's students, Javier Hernandez and Mohammed Hoque, installed a project around the MIT campus called the MoodMeter: a large-scale, long-term smile monitoring system. They positioned cameras in six public places and projected a big barometer that showed the aggregate "mood" of each place. To make the system more legible and interactive, in the cameras' live feeds an emoji face was superimposed on the face of each person who passed: a smile, a neutral expression, or a frown, depending on how the system interpreted their feelings. (This also helped with privacy concerns.)

Mood Meter Project: a large-scale, long-term study to understand mood around MIT.

I was curious to see if, say, Media Lab students were more stressed than computer science students, or if people were happier in the afternoons, or if exam times promoted depression. Nope—none of these hypotheses proved true! As it turned out, the only thing that influenced mood was the number of people in any particular space: the more people gathered, the happier they were, in aggregate. It's a reason to put yourself in social situations every day, especially in dark winters or if you're feeling depressed. (It's also why we all started feeling so out of sorts during 2020's period of isolation.)

Our built environment today is already saturated with *dumb* security cameras. These streams will soon gain cognitive computing skills, including emotional analytics to help retailers, cafés, movie theaters, bars, and workspaces measure activity and affect. Just as a standup comedian gauges audience laughter to curate a monologue, emotional analytics provides an authentic feedback loop for designers of social settings. The result will be more emotionally engaging spaces. Free piña coladas at the airport? (Yes!) Dim the lighting in the restaurant and offer live music? (Yes!) Provide guides to help us shop at the grocery store? (Maybe, depending on the customer.) Using mood as an endpoint, companies will rapidly learn from these interventions, personalize based on specific sets of customers, and have the evidence to justify investments to the C-suite.

Should this emotional analytics data be available to the public, perhaps published as scores reflected on Yelp or Tripadvisor? These unconscious reactions would certainly be more genuine and representative than today's highly subsampled, written reviews, and would be reported faster, even continuously. If you're organizing an eight-year-old's birthday party, you could find out which venue provides higher delight per dollar for third-graders: Build-A-Bear, Chuck E. Cheese,

"AI coaches need emotional intelligence and empathy to understand and support you."

—Rana el Kaliouby, CEO, Affectiva

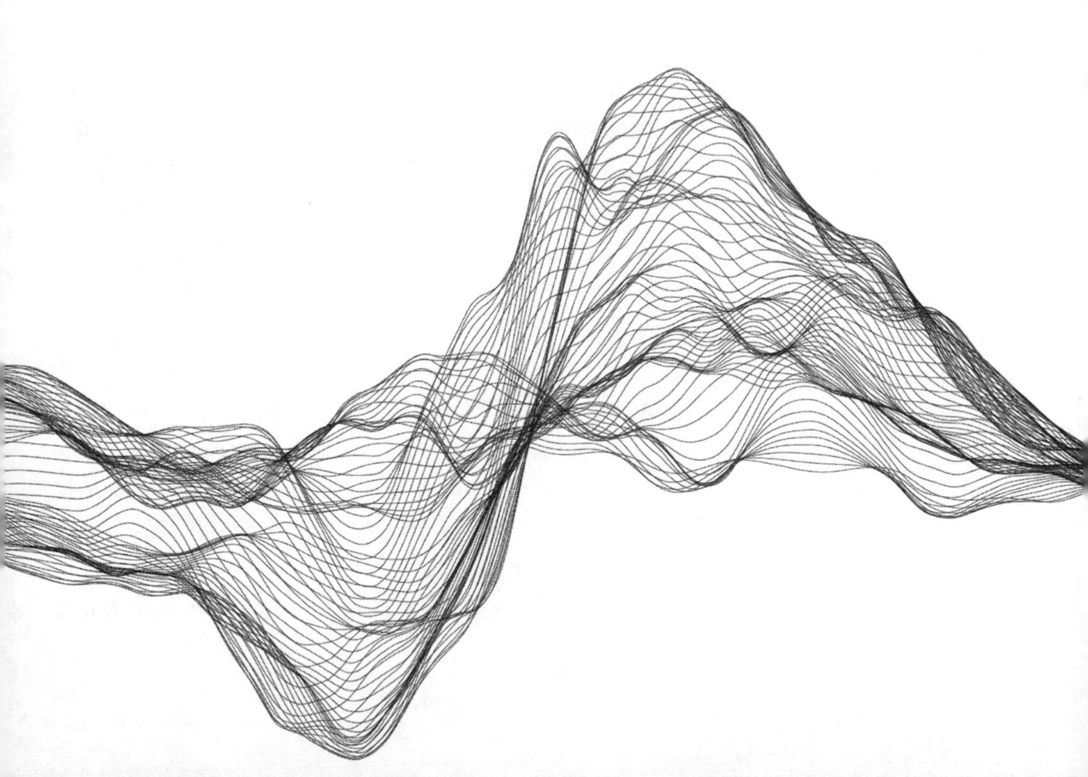

or Jump On In. What if you factor in the mood of the highly-sensitive-to-sound adults?

My money is on Build-A-Bear. Malls, generally, are depressing and soul sucking. They power the hedonic treadmill of shopping as entertainment and promote a hyper-consumerist culture of overconsumption and waste. Yet, there *are* exceptions. In 1999 Joseph Pine introduced the notion of the "experience economy" in his book by the same name, which serves as an inspiration for brands to stand out against the endless aisles and racks of inert, product-browsing department stores like Kohl's or JCPenney, may she rest in peace. According to Pine, stores must differentiate through emotional, engaging experiences: think Nike's running treadmills, or Bass Pro Shop's giant fish tank, where you cast and test lures on real fish (not stunt fish).

At Build-A-Bear, instead of picking a generic teddy off the shelves, you design one yourself. From the fur to the eye color to the level of stuffing—jolly and chubby, or lean and athletic—each child drives their own experience. As a result, it's intimate and personal, and kids have a blast—something you can measure with SuperSight.

I visited Build-A-Bear in the Minneapolis Mall of America just after they rolled out a partnership with Disney to make your own Baby Yoda. The place was bonkers. Kids were running in every direction. Shouts of glee rang across the store to friends and parents. I was reminded of meeting the CEO of Build-A-Bear a few years before at Disney, where we were both participating in a two-day brainstorming session about designing *Avatar*-like worlds. As he talked about the process of designing and refining the Build-A-Bear experience, he said that they would position an observer with a clipboard at each bear-building station to understand how long each step in the process took and come up with ideas to improve it. Ultimately, their goal was to "optimize delight."

My presentation was right after his Build-A-Bear story. I was speaking about how to design enchanted objects—"optimizing delight" is my goal, too. As I thought about the trend to embed sensors, computation, connectivity, and cameras into more and more everyday objects, I had a little epiphany: What if SuperSight could measure automatically the same things that those Build-A-Bear observers were observing manually in their stores? Perhaps computer vision could help track not only concrete measures like time spent at each station, but also qualitative responses—a child's smile, a parent's frustration—in a quantitative way. In my notes that day, I wrote, "SMILE METER service." The idea was to build an emotional-delight feedback mechanism that could look for and optimize visitors' emotional state, and adapt the environment or customer interaction style to people dynamically.

That was in 2017. Now, camera-based analytics are standard in most stores to track the number of visitors, how many visitors touch which products, whether visitors interact with store associates, and the conversion rate from browsers to touchers to conversers to buyers. On top of that, emotional analytics, especially those that measure engagement, have become a hot topic in educational settings. In 2019, the AI pioneer and former Google China CEO Kai-Fu Lee presented a talk at MIT on AI Superpowers, in which he showed a classroom in China where computer vision measured the engagement score of each student. Individual attention and measuring of this sort feels overreaching, paternalistic, and creepy. But providing this engagement feedback to the teacher *in aggregate*, rather than associated with individuals, may help them improve their craft.

In the same way, mood meters will help retailers, restaurants, hotels, airlines, and cruise ships better gauge their customers' experiences and iteratively improve. Could this affective measurement be done manually by positioning people to observe and log this data, as Build-A-Bear did? Yes, but it would be enormously expensive, impractical, and biased.

What will be the secondary effects if mood meters are installed across all the branded venues of the world? The main one will be more people walking around whose primary job is to make you happier (unless the retailer sells twice as much to unhappy people—which is very likely). Think Julie the cruise director from *Love Boat*, whose only job is perfect introductions and social lubrication—or the role that hype men play at concerts. These people are there to inject enthusiasm and energy into events, getting everyone out of their seats and dancing. Perhaps in the future, while robots pour your drinks, the employees formerly known as bartenders will listen to your story, make introductions, and get you dancing on the bar after 11, instead!

Restaurants and retail aren't the only contexts in which a higher emotional quotient, or EQ, would be helpful. Cameras' ability to read emotion also allows for another kind of SuperSight angel on our shoulder: an interpersonal coach.

2.4 Dating dystopia

interpersonal coaches *telepresence crowds*

In the *Black Mirror* episode "White Christmas," Jon Hamm offers dating advice to someone less experienced by seeing through their eyes and whispering in their ear. While the nervous bachelor tries to pick someone up, Hamm's character provides second-by-second instruction: "Don't give her your attention; talk to her friend instead." The person of interest is none the wiser—and neither of them is aware that it's not just Hamm watching, but a whole audience. Dozens of other people (they appear to be all guys) have also tuned in to discuss every move as it happens, like a Twitch gaming livestream. It's super creepy and privacy invasive: some sort of vicarious dating-as-entertainment (or education, depending on how you see it).

This is an extra-dystopian example of personal coaches—thank you, *Black Mirror*. But take out the whole are-you-being-watched thing (but maybe not the Jon Hamm part), and consider all the situations where you might like a more experienced coach (or Jon Hamm) to help you act, react, understand, and strategize about an intimate personal exchange.

We've already introduced the idea of an assistant that remembers names at a conference or party, and even suggests topics or anecdotes to shepherd a conversation. What if they could also suggest ways to deepen your relationship? Newscasters have producers in their ear; why wouldn't everyone want an omniscient coach to help interpret a challenging conversation with a coworker, provide an ego-boosting affirmation before a presentation, or just remind us how to be a more empathetic and consistent parent and partner?

SuperSight makes this possible by seeing in both directions: out into the world, to recognize what is happening around us, including how others perceive us, and also back at ourselves, to register our own reactions and emotions.

What we fixate on, and for how long, reveals a lot about ourselves, and companies are beginning to use computer vision to monitor our eyes' movement patterns to gain insight into our subconscious mind throughout our daily lives. At the 2019 Augmented World Expo, one of these companies shared a demonstration analyzing subconscious eye-gaze patterns to help car manufacturers design more attractive vehicles. By measuring the millisecond dwell times of different shoppers wearing an AR headset, the system determined what features different types of people noticed or glossed over—information companies could use to redesign or rewrite marketing materials to highlight the feature most appealing to their target market. Research also shows that our eyes move in characteristic and machine-learnable ways when we experience various emotions. Social coaching applications will use these to gain insight into users' attention and moods, then personalize interactions far more effectively, offering in-the-moment assistance in many areas of our lives.

Cameras in your glasses will track your gaze, and cameras in your environment will analyze your attention and emotional state. Combine this with data being collected from *others'* glasses, which are looking at you, and you can create a powerful layer of behavioral feedback on every action and interaction throughout your day. I'm sure the interpersonal coach in my glasses would frequently say, "Slow down. Just listen for another minute. Reflect on what they just said with support. Keep focusing on them." "Uncross your arms, relax, and smile." "Ask more Socratic questions rather than reciting parables." (At which point . . . I would fire this AI.)

One of the biggest design challenges with SuperSight is how to expose the data it collects to us—and our peers, employers, and health insurers—in ways that we can consume, tolerate, and act on, ultimately creating positive change. What can we show people that will help them make better decisions for themselves? And with whom will we be comfortable sharing that data? These SuperSight AI services will most likely be subsidized by your employer and health insurer, just as wellness programs are today. This makes data policies and agency concerns more challenging to design—and privacy harder to ensure.

You want to see your own improvement over time, so longitudinal trends and the persistent data storage required to glean these patterns are essential. But who else might get subscription rights to your personal record? Do you opt only to share data with your therapist? What about a curated view for a professional coach who is compensated by your employer? Do you grant the AI coaching agency "anonymized" access to improve their service for others, becoming a willing guinea pig for others' sake? If you change employers, can you transfer data to the next employer, or do you have the right to delete it? And is that negotiation part of the interview process—"Please *show* us a time that you showed leadership qualities"—or only after you've been hired? Imagine an HR service that calls your AI coach instead of your former boss for reference checks. That would revolutionize recruiting with radical transparency—and strike fear into nearly everyone! The "quantified self" is only attractive if you own your data and control access rights.

The multiplicity of SuperSight-enabled coaches we collect won't just help us enliven conversations—they'll contextually teach and guide us throughout our lives. And they'll all use SuperSight to reveal patterns and opportunities for improvement. Wellbeing coaches will observe how we spend our time, when and what we eat, if and how much we exercise, and when we show signs of depression or manic behavior; they'll monitor media consumption, sleep patterns, and the quality of our social web. Omniscient and clairvoyant AI coaches will do the jobs expert assistants do today, for those rich enough to have them: prioritize tasks, populate your calendar, filter communication streams (including the news cycle), and get your body to the right place at the right time. There will be branded coaches from prioritization gurus like *7 Habits of Highly Effective People* author Steven Covey. The more recent GTD (Getting Things Done) religion will offer another. Marie Kondo's app will auto-suggest things in your line of sight that do not spark joy to help declutter your life. And many companies will foot the bill for coaching services, democratizing now-pricey human leadership coaches, because of the lift in your clarity and productivity.

One last example of the potential power of emotional coaching: because SuperSight can interpret others' emotions, it could also help amplify subtle emotional cues for those who need them. Many emotionally myopic men would find this useful, including most MIT faculty and me. But it might be particularly helpful for those on the autism spectrum. When Google Glass came out in 2013, neuroscientist Vivienne Ming saw the potential to help autistic kids like her own interact with the world around them more easily. So she did what any other self-described "mad scientist" would do: created "SuperGlass," an app to classify facial expressions. Stanford researchers tested the system and found it increased autistic kids' ability to correctly identify different emotions, even when not wearing the glasses. Ming's own team found that it helped foster empathy. Of course, Ming wasn't aiming to "cure" her son's autism. As she writes, "I didn't want . . . to lose him and his wonderful differences. SuperGlass became a tool to translate between his experience and us neurotypicals . . . It didn't level the playing field—it just gave him a different bat to play with."

This is the way we should think about interpersonal coaches: not as augmented instruction manuals, or cures, but as tools we can use to help increase understanding and social cohesion. They shouldn't prescribe conversations or behaviors so much as prompts and cues to form better habits and richer connections with each other.

What if you could trust your mirror's fashion advice?

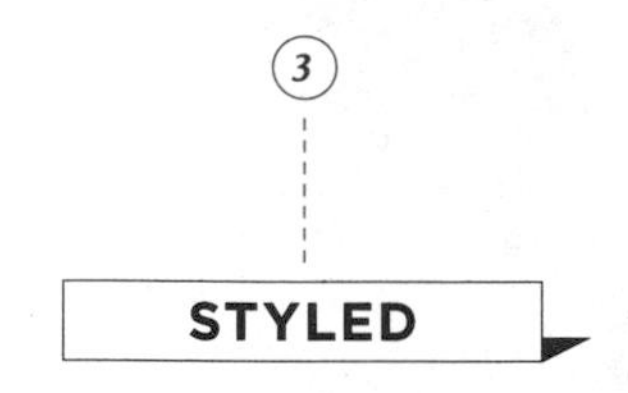

About personal expression, home design, and the future of shopping

3.0 Self(ie) expression

Snap filters *avatars* *Animoji*

Humans have been obsessed with modifying our appearance for rituals or warfare since long before mirrors were invented. Masks, face paint, or adornments of flowers bent our identity toward nobility, bravery, or purity. Seeing ourselves modified or remade is an age-old desire, and in the face-filtered future world of SuperSight, it becomes more commonplace, instantaneous, and mercurial.

New technologies that let us see ourselves better have always sparked revolutions in vanity. First, the hunter Narcissus fell in love with the man reflected to him in the pool, failing to realize that, yes, that hunk was him. Then we polished copper and obsidian stone to get a dim, hazy reflection. The first metallic mirrors were small, expensive, and poisonous; the oldest ones from the first century AD in Lebanon and Venice were made from a dangerous amalgam of tin and mercury. Then silvering was invented, and we began to build furniture around this chemical technology so that we could sit and modify our appearance with paints and powders, unaware that many of these, too, were toxic. The better we could see ourselves, the more interested we became in controlling what was reflected back at us.

Selfie culture provides even more opportunity and expectation, as well as new tools, to craft identities that we project to an ever-larger audience. In her book *Second Self,* psychotherapist and MIT professor Sherry Turkle documents how kids use online personas to experiment with identity and signal affinity to a tribe—or reject one. The self-prototypes of teenagers are often taken to

Fashion is about to break into self-expression extremes due to SuperSight.

extremes: goth clothing and makeup, nose piercings, and tattoos defy belonging to their parents' tribe or express new freedom from authority. As adults, we're often more subtle, but the signaling persists. Daily fashion choices indicate belonging: a well-cut blazer that shows an intended trajectory in some professional hierarchy; a shade of peach blush that demonstrates you know which color palettes are "in" this season; a baseball cap with your loyalty to a school, sports mascot, or brand. Or my favorite, the "no brand" brand, which is itself an indication of tribalism.

In this way, we're already augmenting reality with our sartorial choices. We shave our heads to feel edgy, wax our bodies to feel sleek, or buy high-performance outerwear to feel adventurous. We spend hundreds (or thousands) of dollars a year not just on clothes and shoes and hair products, but on gyms and protein shakes and cosmetic surgery and Noom. We alter our bodies and what we decorate them with, permanently or temporarily, to match an internal, non-visible image of ourselves—or an external expectation imposed by society. We're continually creating identities and signaling status, all through what we choose to look like.

SuperSight will forever alter this universal cultural phenomenon in one significant way: it will make the decisions about the way we augment our bodies not only temporary, but also instantaneous—and perhaps also customized to each new context. This may be the height of identity distortion: to look different to each person who looks.

The first step toward styled SuperSight is the face. Instagram and Snap make it easy and fun for us to augment our appearances, and our chosen identity, in less than a second, then broadcast the result to hundreds of friends, friends of friends, and strangers, depending on the platform. We are literally trying on new versions of ourselves every time we make our eyes sparkle, smooth out our skin tone, or barf rainbows. And if you're a teenager, you're doing it fifteen or twenty times a day.

Soon, instead of just being broadcast via modified photo, your mod will be seen by anyone who sees your face IRL. Your filtered self will appear in any photos that others take, and everyone wearing AR glasses would see your anime-sized eyes, green skin tone, or mohawked hair. Likewise, as you pass augmented surfaces like windows and mirrors, you will also see the lime-green anime avatar reflected back.

In a world wallpapered with screens and digital mirrors, the party of filtered faces will feel like a masquerade ball where the fun comes from just barely recognizing people you know. You'll be able to pick from wildly expressive wardrobe filters every day. Instead of wearing constricting pants or having to buy expensive designer clothes, you could leave the house in your pajamas every day, because clothing is projected, too. As your mood shifts, how you appear will change, either amplifying your vitriol or softening to appear less harsh.

This will inevitably lead to more experimental looks. Instead of committing IRL to a particular presentation, you can just flip on your filter for your friends. So why not change eye color, hair highlights, and eyeglass frames every few hours? We may enter a *Hunger Games*–esque world where we are constantly trying to outdress each other: more sparkles, more wafting lofted wigs, more animated top hats, more iridescent orbiting butterflies, more more more. Designs will get wilder and wilder: animal appendages, chameleon-like color morphing that responds to your environment, epic backdrops like the Eiffel Tower or a tornado cyclone that drops in behind you. There will likely also be a countermovement for those who don't want to partake in this consumer culture—a competition for the most bland, the most simple.

Wardrobe will also be relentlessly dynamic, changing throughout the day based on a feedback loop of actions and emotions, both yours and others'. When I walk into an IRL meeting and no one else is wearing a tie, I discreetly remove mine. With SuperSight, instead of having to observe what other people are wearing and making a decision based on that, your glasses will adjust your suit to suit. You might be prompted to tone it down if you're getting too much attention. Or if no one is looking at you *at all*, maybe it's time to put on something more eye grabbing, like a kinetic turbine hat?

To test the theory that we are trending toward a hyper-expressive future, I met up in Paris with Ian Rogers, the chief digital officer at the mother of all fashion houses: LVMH, the parent company of Louis Vuitton, Fendi, Givenchy, Dior, and seventy-five other luxury brands. Rogers started the conversation about our augmented fashion futures by stating that "AR is going to change our lives, especially in the zone of fashion and identity." As their team tracks social media, they are seeing a "massive upward trajectory of tattoos, both virtual and real," as well as other personal-expression signifiers. "We want unique identities and status—but this differentiation is nested in brands that are part of our culture." As we become more comfortable with digital augmentation, we'll become more willing to manifest those identities in physical, analog ways, too. We may start changing clothes in the middle of the day just for the sake of it, or three times during a party, Madonna style—which will of course please fashion retailers' bottom lines. "Renting makes you more daring," Ian notes. "People go out of their comfort zone and try new identities."

SuperSight will forever shift how we express ourselves—and how we are seen by others. As our virtual identities become as fluid as the mercury once used to make mirrors, who will we choose to become? And how will the new imagination engine of SuperSight affect the pace of fashion, and the experience of consumption?

3.1 Mirror, mirror on the wall

fairy tales *rethinking retail*

"Mirror, mirror on the wall, who is the fairest of them all?"

When the queen in the fairy tale *Snow White* asks her narcissistic question, she is looking for reassurance of her beauty. Instead, she is met with an unwelcome judgment. The AI camera embedded in her "magic" mirror takes a moment to calculate her feature vectors and wirelessly upload them to the cloud. A neural network then instantly cross-checks her visage against all faces in the kingdom. The result lands hard for the evil queen that night (and puts a pretty young lady in a nonenviable near-death situation with seven mournful dwarves).

What if we really could consult a talking mirror in our homes? A mirror that offers style advice more honest than any well-meaning mother's, and more objective? Would we want it to eschew assurances of *You look GREAT!* in favor of reality, or would we want it to espouse confidence so we continue living in that well-lit bubble?

As we discussed in the last chapter, we're beginning to see magic mirrors that facilitate our home workout routines. But technologists have been flirting with other practical (and vain) mirror applications for some time.

At a 2004 TEDMED conference, I presented an augmented mirror, designed to be installed in your bathroom for a daily glance, that superimposed health biometrics like weight, recent physical activity, heart rate variability, and blood pressure onto your reflection and highlighted trends or reasons for concern. Superimposing information on the body—pulse and blood pressure shown on the heart, stress levels on the forehead, activity over arms and legs—was motivating. It made daily health feedback unavoidable, and seemed to psychologically involve people in a promising way. Now we can do so much more with home mirrors beyond health and wellness feedback and glanceable weather, including fashion projections.

Years later, I met a master's student at MIT's Sloan School, Salvador Nissi Vilcovsky, on whose project I agreed to advise. He came from the fashion world and had a vision for a new type of try-on mirror at high-end stores like Prada, Dior, Armani, and Tom Ford. With a strong engineering team in Tel Aviv, we created a magical mirror called MemoMi. It recognizes you, allows you to compare outfits, and suggests other items you might like. Customers try on one outfit, then twirl in front of it so the mirror can take a 360-degree video. As you try on multiple outfits, the mirror shows them side by side so you can see which looks best. The next time you come in, the mirror remembers you and your purchases to get better at suggesting items to try on.

Salva installed the first version of MemoMi at Neiman Marcus in 2015 to much acclaim. Not only did people enjoy the experience, but they often sent their side-by-side spinning photos to friends and family to get their input and feedback. This also made them more likely to be confident in their purchases.

The return rate on these items plummeted from an industry norm of around 25% to less than 15%, unexpectedly solving a persistent problem in the apparel industry.

After the president of Uniqlo saw a demo at the National Fashion Retailer Big Show, he asked Salva and MemoMi to develop an AI mirror for the Japanese chain. Uniqlo is a brand of many colors, so this version of the mirror had a new algorithm that recognized the edges of sweaters, pants, or dresses, isolated an item of clothing from the background, and then dynamically changed its color. Instead of trying on four different versions, you just make a swiping gesture in front of the mirror to change its color. (BTW, I am not pretty in pink; I learned to stick with navy and black.)

But how do you objectively know if what you've chosen looks any good? Whether a clothing item "fits" you is a subjective judgment that no one thought could be made with traditional computer vision. But neural networks trained on tens of thousands of positively and negatively scored examples of too-tight, too-loose, and just-right fits can start to render more accurate opinions. Add to that mirrors that know the exact contours of your body—like Naked Labs' model,

The MemoMi mirror helps you compare outfits side by side or change the color of your dress.

which uses infrared projection to take an array of measurements of your body and track changes over time—and you will feel confident that you're walking around in well-trimmed style.

As these mirrors further refine their ability to see and reason, companies will provide new services that revolutionize fashion, apparel, and beauty, spurring billions in bespoke product orders. In addition to lining the walls of stores' dressing rooms, these mirrors will be installed in homes and high-end hotel rooms, where they will 3D-scan our bodies, compare that information with cognitive computer services, and make sophisticated and timely fashion recommendations.

You'll no longer have to ponder what to wear as part of your morning routine. Taking stock of the clothes in your closet and factoring in the latest style trends, the day's weather, that big presentation on your calendar, and knowledge of what your boss will be wearing—she has an enchanted mirror, too—your mirror will recommend three excellent options.

In this way, the presence of enchanted mirrors will transform how people shop for clothing, shoes, accessories, and certainly makeup. You'll download mirror-enabled apps from fashion labels or celebrities and receive branded style advice and shopping recommendations tailored precisely to your unique body shape, tastes, and needs—H&M's "Summer Vacation Lineup," for instance, or Beyoncé's "Work It Out" gym gear (which will, of course, include some Ivy Park). The business model for these recommender services is revenue share: the app takes a cut of the money you spend when you purchase something it recommends.

A new era of influencers is coming to your dressing room. Bloggers and Instagram stars will not only give you static commentary on the latest trends via online videos, they'll also appear in your mirror with some honest early-morning advice on what you're planning to wear. The YouTube makeup tutorial stars of yesteryear will sell packages of online mirror courses to help train budding kings and queens of bronzer. Or you could let Fashion Institute of Technology students help you color-match.

To truly thrive, fashion brands must begin rethinking traditional channel strategies. Dedicated flagship stores are expensive and visited far less than gyms or corner cafés. How might brands deploy enchanted mirrors to deliver virtual

"We're already augmenting reality with our fashion choices. We're continually creating identities and signaling status, all through what we choose to look like. SuperSight will make these decisions not only temporary, but also instantaneous—and customized to each context."

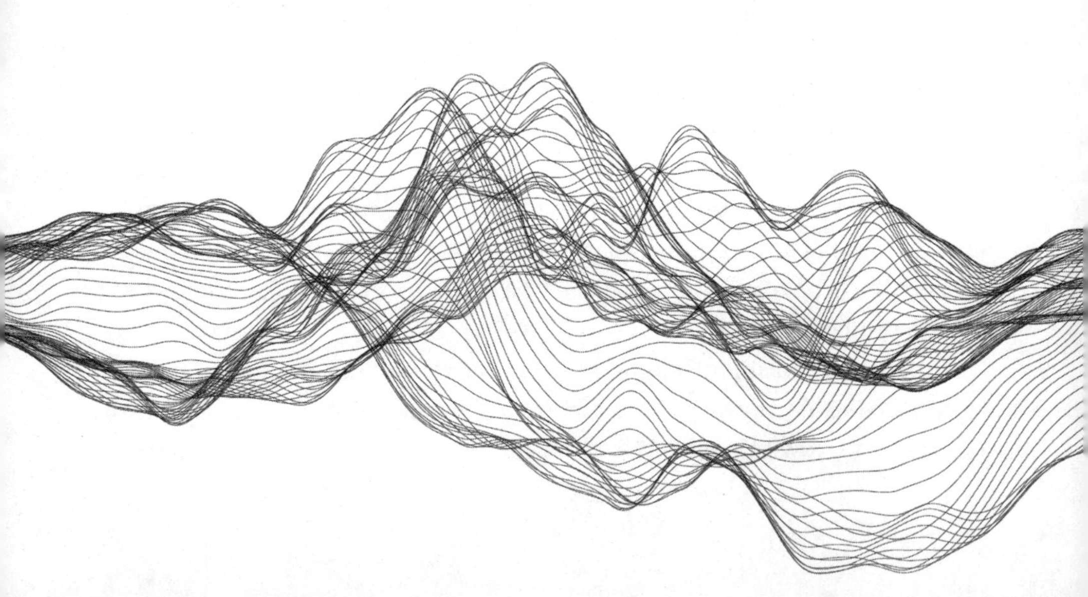

try-on experiences infused into these and other higher-traffic venues: hotel suites, restaurants, concert hall lobbies, and homes? If the big, storied fashion brands are slow to embrace this ambient-try-on experience, then it'll be the well-heeled startups that will be the fairest of them all.

New mirror technology will make it easier to access and try on options when it comes to personal expression. The downside of that, of course, is the number of options itself.

3.2 The paradox of choice

virtual try-ons *solving fit*

And you thought shopping for salad dressing or vitamins was overwhelming.

Walk into an eyeglass store and you are hit in the face—almost literally—with the number of frame styles available. Every possible material, shape, size, and brand is calling out for you to try them on. And, chances are, 90% are going to look *terrible* on you. Some swamp your face; others get lost in the sea of your eyebrows. Classic Ray-Bans just make you look like your grandfather. And aviator frames? You're no Maverick.

A pair of specs says a lot about your personality and style; which to buy is a big decision. After all, they will be sitting on your face, front and center in every conversation you're going to have with anyone in the world for the next couple of years. Yet how do you begin to boil down the ocean of options and make the right selection?

Barry Schwartz, a behavioral economist from the Wharton School, describes what he calls the *paradox of choice* in his landmark book of the same name. The book collects studies on consumer behavior and concludes that shoppers would be much happier with *less* choice, not more. His famous salad-dressing study proved consumers were more likely to buy if they were presented with fewer options. Abundance of choice confounds us. But SuperSight solves the paradox of choice by reducing the complexity of browsing very large product assortments, while still providing great fit and a very personal style.

When I was at Warby Parker, my team developed a virtual try-on (VTO) service that embeds a retail experience in your phone, guided by SuperSight.

Thanks to the very sensitive 3D camera mounted on the new Apple iPhone X and beyond, we were able to cast forty thousand infrared dots onto the contours of the customer's face and read them in less than a second. Then, we mapped these TrueDepth results to a topology of key facial features (pupillary distance, nose bridge height, cheekbones, etc.) to recommend a subset of glasses most likely to fit their face. In other words, the same technology that allows your face to unlock your phone can also help alleviate the paradox-of-choice shopping problem.

To optimize VTO's recommendations, we first gathered data about which glasses Warby Parker's existing customers had purchased, and which they'd tried on but hadn't purchased. Since Warby sends customers free boxes of five frames to try on at home, this training data was readily available. We experimented with an ensemble of neural networks, too, each with a different job. The first optimized face shape to glasses shape, the second scored the fit between

In less than a second, Warby Parker's find-your-fit tool scans the contours of your face, then recommends a set of glasses most likely to fit. We developed this measurement-based recommender tool a year before launching the virtual try-on visualization.

glasses and your head, and the third evaluated the tonality of your complexion to match glasses materials. Taken together, this ensemble of AI-based judgments allowed us not only to offer great style advice, but also to tell someone *why* a pair of glasses looks good on them—or not. (Using an ensemble approach with multiple neural nets solves the *explainability problem* in AI: the challenge of helping humans understand machine learning algorithms that are otherwise a black box.)

Warby Parker's award-winning virtual try-on. Here I'm wearing real glasses and projecting the virtual crystal version of the same frames to show the quality of alignment.

When we launched, VTO was a sensation. It was the first real success story of augmented reality for ecommerce. Warby Parker won a 2019 Webby Award in the fashion and beauty category, and the company saw an immediate return on its investment. VTO also brought in many new and different types of customers, including tech enthusiasts and people outside urban centers who had never been to one of Warby Parker's physical stores. We saw a big increase in app downloads, with millions of people trying on glasses every month. Retail associates reported increased customer confidence: people walked into stores with a short list of products to touch and feel, were primed to purchase more quickly and with more assurance, and made fewer returns. Most surprising, though, was that, for the first time, many people were ordering glasses directly from the app—*without ever physically trying them on*—rather than signing up for Warby's home-try-on service. Buying a $20 T-shirt sight unseen from Instagram is one thing; a prescription purchase like a pair of glasses that you wear every day, which come to define your identity for years, is quite another.

As Warby Parker grew, a big focus was building more physical retail locations across hundreds of US cities in the highest-traffic, highest-rent locations like Rockefeller Center in NYC. Because these stores have millions of potential customers walking by their windows after business hours, I thought we should make a special window display that incorporated VTO. So, using the same face-scanning, data-mining, and augmented projecting technology developed for the app, we created a window display that projects five different pairs of perfectly fitting glasses onto your face as you walk by the store. With each step, you could look over and see a different set of frames in the mirror display of the store window—all of which fit, because the system 3D-scans your pupillary distance and face width, and matches your complexion and face shape.

This spatialized, glanceable, "ambient-commerce" try-on experience could soon be used in more shop windows, especially for products that can be projected onto your body. Millions of people pass Cartier stores on Fifth Avenue over the course of a month, where managers foil theft by removing the brooches and necklaces that typically bedazzle the fake busts. Yet with AR, the store could project their sparkling catalog after hours onto every person that passes by!

Many industries outside of eyewear are also addressing fit problems. How many times have you taken something off the rack, only to find it too small in the chest? Or ordered basically anything from the internet? We're incredibly bad at guessing how we'll look in the clothes we buy, an expensive problem for companies. According to the National Retail Federation, Americans returned $260 billion in merchandise in 2019, or 8% of all purchases. Ian Rogers, the LVMH head I mentioned earlier, confessed that luxury brands like Dior, Fendi, and Gucci suffer from a 30% to 50% return rate for their products. Getting the fit of clothing right is tough, which is why we often order multiple sizes—especially when we aren't the ones paying for returns.

The environmental impact of all this returning is significant and unsustainable. For each return, you not only have to factor in the miles it has to travel to a processing center, but also its being industrially rewashed, repackaged in new single-use plastic, and then shipped out again—if it even gets repurchased. Ten percent of online returns in the US get incinerated or donated, and the remaining 5 *billion* tons of fabric end up in landfills every year. VTO-like tools that scan other parts of our bodies—feet for shoes, wrists for watches, fingers for rings—will help customers make better-informed fit and style choices, reducing waste and returns down the line. Everyone wins! Including the environment.

ZoZo Suit tries to solve the "fit problem" with computer vision by sending you a polka-dotted unitard. You spin in front of your camera as they digitize your dimensions.

ZoZo is among the leaders of this movement to minimize returns with SuperSight tech. It was founded by Japanese entrepreneur Yusaku Maezawa, who became a billionaire with Zozotown, a fashion website catering to the young and fashion forward that dominates Japan's online clothing market. To solve the fit problem, they've created a skin-tight unitard covered in what look like Ping-Pong balls, called a *fiducial suit*. It's similar to the suit worn by Andy Serkis as he hopped around the set of the Lord of the Rings films pretending to be Gollum. The computer vision–friendly garment helps build a very precise digital rendering of your body with all your measurements—then uses that profile to sell you clothes.

Imagine if that body scan could ensure every single thing you order fits you perfectly. What a dream! And you don't even have to go to a store—ZoZo sends you the outfit for free. Put it on, set up your phone, and rotate in 30-degree increments until you've captured all 360 degrees. The system now has a 3D model of your body, and your fit problem—plus the retailer's return-rate problem—is solved.

By empowering customers with new tools for precise measurements, we can limit waste, save retailers and manufacturers lots of money, and streamline the shopping experience for consumers. And when everything you own looks so darn good on you, you're probably going to buy more. Now—where are you going to store it?

3.3 Predictive wardrobes

subjective neural networks

At the Copenhagen Interaction Design Institute, I lead a yearly workshop on enchanted objects with a magician from London named Adrian Westaway. After students study the history of magic and learn a couple of tricks, Adrian and I give students an assignment: go to IKEA and prototype a piece of enchanted furniture. Students start with ordinary items—curtains, coffee tables, lamps, and cupboards—then deploy the tropes of magic to create novel products that satisfy essential human needs and desires.

As you would expect, the students prototype wonderful objects. Some of my favorites over the years have included magic mirrors that show your future self, given your current smoking, exercise, and bad food habits; robotic lamps that infer your current activity (reading on the couch, running around cleaning, dozing off) and adjust lighting levels and tonality to accommodate; and Amazon-linked kitchen cabinetry that automatically refills itself like the minibar in a hotel room, then charges you when you pull out the extra dishwashing soap.

One provocative idea, made possible by knowing fit and applying big-data predictive algorithms, was to proactively stock clothes closets. Instead of suffering at the shopping mall or ordering the wrong size or color of something online, your wardrobe would autofill with outfits just for you, based on your measurements and what you've worn and liked in the past. This *Clueless*-esque concept radically pivots the business model and process of shopping toward a burgeoning trend: the predictive economy.

The vanguard of this trend, digital personal shoppers like Trunk Club and Stitch Fix, are doing incredibly well in the market; before IPOing in 2017, Stitch Fix had generated nearly a billion dollars in revenue despite obtaining less than $50 million in funding. These services try to understand your fashion preferences, then proactively ship you clothes that you might want to purchase.

Here's how it works in the case of Stitch Fix. First, you take a personalized style quiz where you give the system basic information like sizes, preferred colors, and desired formality level (do you want clothes for work or play?), then let it know a couple of things you love and hate, like showing off certain body parts or disguising others. You also rate some outfits they show you by giving them a thumbs up or down. Pleats are too '80s—but the all-black biker look? Love it. They then send your five items based on this test. You try them on, and only pay for those you keep (minus the upfront $20 styling fee).

As Stitch Fix matured, it began "weaving data science into the fabric" of the company (a punny tagline on their digital Algorithms Tour too good not to shout out). Every piece of clothing is tagged with about a dozen categories. Stitch Fix still uses your style test to begin narrowing down what it has in its stock room, but now you're not the only data point: it also factors in what's trendy, and what other people with similar profiles to yours have loved, among other criteria. Every time you send clothing back, you're helping train not just your own personal algorithm (so each new box of clothes is increasingly likely to be a home run), but other people's, too. It's great for you, and great for Stitch Fix, because you're buying four out of five items at a time, versus just one or two.

SuperSight—especially when paired with the magic mirrors I've described—will fuel a future where clothes always fit because your body dimensions are precisely known, your style preferences are recognized from what you typically wear to various events, and your aspirational fashion ideals can be determined from the influencers you follow. These predictive systems will get so smart at anticipating your needs that shopping will mean simply opening your closet, plucking out the next outfit, and putting it on.

Services like Stitch Fix currently ship you a box. But the wardrobe I mentioned earlier, which receives clothing deliveries automatically, and without packaging waste? Personally, I would love this service; it would help me when I'm scheduled to travel to a talk, I'm behind on laundry, and my plane is leaving in a couple of hours.

Amazon is working on these predictive algorithms too. In fact, I suspect the 2017 launch of the Amazon Echo Look, a service that recommends which outfit looks better on you, started as an experiment to understand the durability and sensitivity of style preferences. After you take photos of yourself wearing various

Echo Look service from Amazon selects your outfit for the day by using a subjective neural network to make a style-based personal recommendation.

outfits, a subjective neural network (one that makes a judgment based on intuition, instinct, taste, and/or feeling—typically only a human trait) scores which one best complements you given current fashion trends. (Subjective nets play to one of the strengths and conundrums of deep learning—their AI logic is opaque: *That outfit just looks better on you. Don't ask me to say why; it just does, for hundreds of reasons.*)

In the Stitch Fix model, you only pay for what you actually wear, and have the option to send ill-fitting items back without paying. Your future wardrobe will work similarly: it will be prestocked with outfits, and SuperSight will facilitate a pay-by-use business model. When the camera in your car or doorbell sees you wearing that new jacket outside the house, your credit card is dinged for its day rate.

The predictive-consumption-with-micropayments model is already being used in a totally different context: predictively stocked hotel minibars. Automatic sensors are tripped whenever you move something, and immediately tally your sins by billing your room (private Pringles binge, be damned). They correctly predict that you will crave and consume alcohol, salty chips, nuts, and chocolate if they are available (and expense-able) at nearly *any* price. You might never go down to the lobby and pay $6 for a Snickers, yet if they're sitting right below the TV . . . As hotel conglomerates get to know you over repeated stays, savvy brands will buy/sell/share guest preference data and personalize what appears in your room; you'll be able to expect a minibar stocked with your fave beers, nuts, and fair-trade chocolate. That wine vintage you've sipped the last two visits at the lobby bar? There will be a bottle waiting there to tempt you.

In fashion, I call this minibar-service model a predictive wardrobe. It fuses the insights of big data, contextual understanding of things like the weather and your calendared events, and a business model that is late binding (meaning there is a bias for providing something of value first, then closing a financial transaction later). There are still plenty of service design issues to figure out: Which service gets how many hangers in your wardrobe to prepopulate the next five or ten outfits? At what frequency will outfits be shipped—daily, weekly, monthly? How will customers prefer price be conveyed, if at all? Like any new product or service concept, these questions will be answered over time through customer interviews, prototypes, testing, and likely some startup failures.

SuperSight also might change how we provide feedback on products, from star ratings to smiles. Cameras in retail environments (or even in SuperSight glasses) will be trained to understand your microexpressions to reveal subconscious reactions to outfits and accessories, as well as how *others* respond to what you're wearing. Using this feedback, the cameras' neural net is retrained for better subsequent dressing recommendations. You may think that you look good in the mirror and give yourself a half smile, but what does your family think? (Like when my teenage daughter smirks and says, "Nice sneakers, Dad," when she thinks they're a little too hip.) Did your colleagues at work remark on your shoes, or more likely just raise an eyebrow? These signals will be automatically aggregated to provide feedback for outfits and inform what gets shipped to your home next.

Ultimately, with predictive wardrobes, we won't need to *own* clothes—we'll just lease them, like millions are doing now through Rent the Runway, a company now valued at more than a billion dollars. Their service allows you to rent items of clothing for short durations or single uses—picture renting that fancy suit for your cousin's wedding in the summer, for 10% of retail, and returning it before it no longer fits. This allows people to stay on trend, without the waste of cycling through cheap fast-fashion (of which an estimated 50% of items get thrown out within the year).

Of course, there are environmental downsides to the rise of the rental economy, too, similar to the problems with clothing returns. Every time you return a piece of clothing to your rental service, it needs to be sent out, industrially washed or dry-cleaned, and repackaged before being delivered to someone new.

(Rent the Runway has the largest dry-cleaning service in the world; it processes two thousand garments an hour.) But returning your rental item is clearly better than throwing it out. And isn't it better to have loved and returned than to never have loved at all?

It's also worth noting: if we don't own our clothes anymore, we won't need to store them, either—our closets will only contain the next outfits recommended to us, so they won't require as large a footprint. Your gym will host your gym clothes, your office will provide multiple work outfits, and your evening attire will arrive at your hotel minutes before your date for a Superman-style quick-change.

We won't need to wash clothes anymore, either: our hamper will become a FedEx drone package that whisks clothes back to a distribution center with associated data that informs what outfits get shipped to our tiny closets next.

Next, let's expand the conversation on these new SuperSight capabilities beyond adornment to interior design.

3.4 The failure of imagination problem

proactive parametric design

As anyone who has ever brought home an armchair half an inch too wide to fit through a door frame knows, the problems we have when it comes to clothes and glasses also apply to interior decoration. For example, paint chips are somehow never enough to accurately gauge the way a color will look on your bathroom walls; what looks like the perfect shade of duck-egg blue on an inch-by-inch square is nearly always too dark or light when applied en masse. Not even the professionals can visualize what colors look like from chips; savvy interior designers will purchase five pints of paint in various colors and actually apply them to the walls of your living room before making a final choice. It's curious. Why can't our brains extrapolate how a simple color will appear?

Here is an even more difficult challenge. Close your eyes and try to imagine the room you are in now devoid of furniture, lighting, curtains, and rugs. Now choose a style—let's say mid-century modern—and slap in a new couch, chair, mirror, area rug, lamp, desk, and some art on the walls. Let's say this redesign was going to cost you $10K. Would you feel confident enough in the accuracy of your imagination to move ahead and order all the component parts? Neither would I.

I frame this general class of problem as *failure of imagination.* Of course we'd like to immediately be able to see what that de Kooning print would look like on our wall, or whether that lawn furniture would match the fence out back, but most of us just don't have the spatial training and creativity to visualize it. But now, with SuperSight, we can. We'll even be able to tell whether that new couch will safely fit through the door—before the delivery guy is swearing in your stairwell.

Consider buying a piece of art. It's a flat, static object that needs to be imagined on a flat, static wall. Seems easy enough, but if your failure to correctly extrapolate how paint in a slightly different shade of gray can accidentally change the whole vibe of a room, what about a large, multicolored artwork?

It's notoriously hard to sell art online. It's simply too hard to look at a screen and have your emotions evoked. It's also tougher to imagine that painting hanging over your mantel—even if you're sitting in front of the fireplace right now. What will the colors look like in real life? How big is 32" × 16" × 20", anyway? ARTSEE uses AR to project a print at the right scale directly on the wall of choice, just above your strange-colored couch and cat-scratching sculpture. Now you can see if that abstract painting is overwhelming or too small, clashes with your existing furniture, or offends your cat (maybe she'd prefer a landscape?). Will it convince you to drop a few thousand dollars on an investment seen only through your phone screen? Maybe, because you (and your cat) have now seen it in context.

What ARTSEE is doing with flat AR isn't technically difficult. It's more challenging, though, to apply patterns and textures on irregular objects like furniture. Visualizing paint swatches at room scale is possible via mobile apps from Sherwin-Williams or Home Depot, but reupholstering a crappy couch with patterned fabrics, or trying out diaphanous, flowing curtains at sunrise, noon, and night, is far more complex. Realistic draping models are, as the rendering experts say, computationally expensive. Fine for a Pixar film, but voluminous window treatments are not coming very soon to a virtually staged Zillow listing near you.

VTO for rigid objects, like furniture, is now broadly available. If you're shopping at IKEA, Wayfair, or Home Depot, you can "land" appropriately scaled

IKEA uses AR to populate your entire room with art and lighting that stylistically matches your furniture and fits in your space.

furniture into your living room, so you can see exactly what that dining table will look like there. IKEA Place's AR app sells hard-to-visualize furniture much faster than traditional showrooms. Context is king for your king bed, lounge chair with ottoman, or long credenza. (And dropping AR couches on subway platforms now seems to be both a wish and trending meme.) IKEA's Studio Mode also interprets your existing space to suggest home furnishings that both fit into a room and complement your current color palette and even style, whether Nordic modern or Victorian shabby chic. And with new, AI-proposed virtual furniture in place, you can use your phone to walk around, check out the proportions, and peer up close at fabric details, then rearrange with the swipe of your finger, no lower-back pain required.

IKEA's AR app includes another important trick: letting people toggle dynamic fixtures to change lighting of virtual scenes. Kaave Pour, director at SPACE10, IKEA's research and design lab, calls this "playful lighting interaction." Like being able to dial ahead in time to see how prints fade, or rugs pill and stain with natural wear and tear, it's the kind of simple, delightful feature that helps rationalize buying more expensive but durable products.

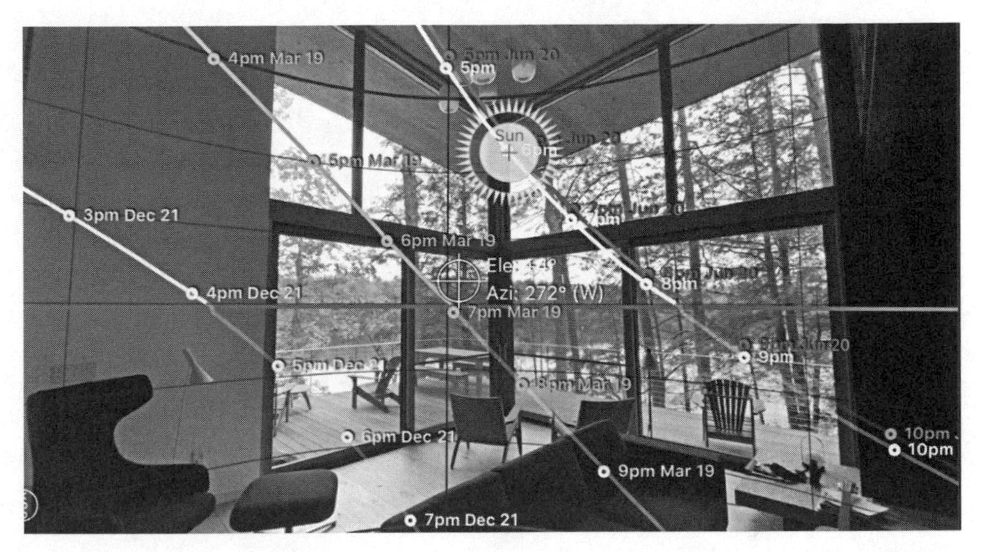

Home shopping for light-lovers with the SunSeeker AR app.

The best light for any home is sunlight. The AR app SunSeeker shows precisely where the sun will be in the sky at any time of day in any season. It locates you with GPS, then projects the paths and position of the sun in the sky above you. It's perfect for home shopping on a cloudy day, anticipating what light comes in from what windows on the vernal equinox . . . and choosing the right spot to place your couch for a nap in the sun!

SuperSight can help people reimagine what is outside, too: new path lighting and darker siding on your house, or protected bike lanes in front of the school. Helping people visualize city-scale changes in context is a big communications challenge for neighborhood associations and towns. When town councils meet to review a worthy proposal for something like bike lanes, it is often vetoed because people can't imagine it, and are therefore hesitant to spend their budget. Static renderings of new buildings or parkscapes don't provide a realistic understanding of walking through an environment, just as that one-inch paint swatch didn't tell you what a terrible idea mustard walls were. SuperSight can provide this in-context visualization to anyone with a smartphone camera.

3.5 Buying with your eyes

social-media shopping | *scene segmentation* | *hedonic treadmill*

Better visualization is not the only way SuperSight makes shopping more immediate and tempting. Though we can already make hasty shopping decisions with a mouse click on a website, and increasingly within Instagram and other social apps, SuperSight will allow us to express interest and "like" with a long glance, and even make a purchase decision with a wink and a nod.

We encounter things all the time that inspire us: clothing and accessories worn by celebs we see on TV, tools used in cooking or home-repair shows, chairs in clubs, fixtures in hotel bathrooms, paintings in lobbies, shoes worn by trendsetters walking past us on a trip to Tokyo. How do we hyperlink these objects of interest to their source so we, too, can be the owners of a bedazzling metallic floor lamp?

As I described in the Introduction, this question interested me so much that I built a company to answer it. In 2015, online photo sharing was exploding. People were posting 300 million photos per day on Facebook alone. Meanwhile, computer vision algorithms out of academia were becoming more capable and performant, and cloud computing was getting affordable enough for seed-stage startups to imagine sifting through these images for interesting signals. Could we train a system to recognize all the specific items in photos, then superimpose links to their source? I teamed up with former Media Lab student Josh Wachman, who more than a decade earlier had created a company called WatchPoint that allowed viewers to select and save products they saw on TV, and fellow entrepreneur and MIT AI-lab wizard Neil Mayle to start the experiment.

Our bold bet about the future of shopping was that people wouldn't primarily discover products in catalogs or stores, but instead by noticing what friends or celebrities wore, ate, or used. Say a high school friend posted a picture of himself skiing with an orange backpack you really liked. Our goal was to connect what was pictured in a photo—a bag, skis, sunglasses—to their provenance or origin, then let users learn more, bookmark, and maybe buy.

We called this "social shopping" service for adding links to social media photos Ditto, because it let you quickly follow a friend's decision. Then came the hard part: trying to label every object and experience pictured in a Facebook photo. We started with brand logos like Nike, Adidas, and North Face, and iconic fabric patterns like Vera Bradley's and Gucci's, spinning up hundreds of cloud servers to aim this massive fire hose of photos into our neural nets. We were paying over $30,000 per month in compute costs just to keep up.

The next major step in enabling social shopping with Ditto was a vision technique called *scene segmentation*. Initially, we could only add linkable bounding boxes on logos and objects; with scene segmentation, we could find exactly which pixels were occupied by which objects in a photo or video.

This opens up a new interaction paradigm where you can "click" on any object in a frame. Being able to turn a word on a webpage into a hyperlink to learn more was the killer feature that created the wonder of the web; to have the same power of interrogation for any object in your field of view was mind blowing. With Ditto, you could not only click to bookmark something or shop for similar items, but also act on each item in an image with specific commands. *Learn* linked to Wikipedia entries for that object. *Go* gave you the Google Maps directions (or Google Trips flight information). *Watch* found the object inside YouTube videos to animate it. *Donate* was useful for objects with causes like parks or WWF cards. *Browse* took you to ecommerce links on eBay, Amazon, or local product sellers where you might buy the item.

Our fascination with the prospect of reinventing retail led us to start Ditto, but our scope was growing. Computer vision plus scene segmentation gave us a new way to interact with the world. Pointing your phone and being able to

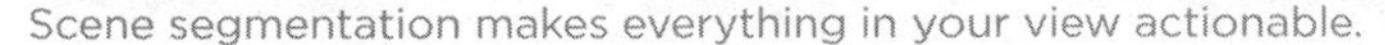

Scene segmentation makes everything in your view actionable.

connect to services was a new visual form of Googling. It was becoming a much larger idea that opened up new deals.

Rather than in a stand-alone app, the Ditto service lived inside Facebook, since that was where all the photo sharing was already happening. A project with Fox Sports meant training the system to recognize every major league sports brand so we could identify team logos worn by users' friends, then superimpose links that let them echo their friends' enthusiasm by getting their own fan gear or deals on tickets to the next game.

Over time we added the ability to recognize things that indicate a lifestyle interest, like snowmobiles, pickup trucks, dogs, and camping gear. After that, we trained up thousands of contexts, allowing Ditto to recognize not just what a photo depicted, but where it was taken. Indoors or outdoors? Inside a café, bar, cathedral, or hotel lobby?

After meeting with some ad agencies, we realized there was another value to Ditto we hadn't considered: sending an organic stream of customer analytics back to brands. Who was wearing their products, drinking their beer, or staying at their hotels? How, when, and where were people using those products? For example, liquor brands were particularly interested in how people were mixing their products. Was ginger beer having a moment right now? Has the pomegranate juice trend died off yet? We started working with Anheuser-Busch, Red Bull, Estée Lauder, Cadillac, and other brands interested in greater insights about who was using their products, when, and how, out in the world.

As the company and technology matured, clients approached us to build custom classifiers to find things that I wouldn't have imagined! A public health think tank wanted to measure cigarette smoking and vaping behavior across US cities revealed in geotagged social media photos. You'd be surprised how many people proudly post public photos of their tattoos, including marks of gang membership. An international beauty products company came to us with a taxonomy of hair types (flat, wavy, semi-curly, very curly, kinky . . .), and with their hundreds of examples of hair-training data, we built a neural network to recommend hair-care products. Now Estée Lauder, Sephora, Clinique, and Kiehl's are training neural networks to sell every category of beauty product based on skin tone, eye color, complexion, inferred age, cheekbone prominence, and more.

We sold Ditto in 2017 to another visual search company, Slyce, which now powers ecommerce sites for Napa, Bed Bath & Beyond, Home Depot, Tommy Hilfiger, and more. Customers just snap a photo to find a matching car part, home good, or garment. It's faster than typing, and works even if you don't know the name for that muffler, faucet, fencepost . . .

The computer vision wave moved faster than any enabling technology that I've witnessed as an entrepreneur. Typically, when you launch a startup after coming out of a strong academic setting like MIT, the reason it fails isn't because you didn't see the future clearly, or because your engineering wasn't sound, but because you misjudged the market timing. You're chronically too early; adoption takes more years or decades than you predicted. But with computer vision, Ditto was at first too early and then, *whoooooosh*, too late. By mid-2017 all of the big cloud companies like Microsoft, Amazon, and Google were madly hiring PhDs, training popular classifiers, and using machine learning as bait to get large companies to "lock in" to an enterprise cloud relationship. Facebook's primary attraction is surveilling your so-called friends, and Google's LENS feature performs much of what we were building toward for visual search. Yet even today, we're still not automatically adding metadata to Facebook, Twitter, Snap, and TikTok videos. Instagram

Ditto those frames. Warby Parker's co-founders posted this photo on Instagram with ecommerce dots so fans could click and buy the glasses.

does allow brands to manually insert shopping dots in a photo, but our vision for a seamless social shopping future is yet to be realized.

Once social shopping is built into smart glasses as I expect, however, it will more than subsidize the hardware and 5G bandwidth costs for every shopper. Viral hedonic treadmill, here we come!

Part 2

Organizational Scale

Computer vision won't just affect individuals and personal interactions. SuperSight is poised to revolutionize some of the largest categories of our economy, like the food industry; improve access to collaboration and learning while boosting their effectiveness and efficiency; and trigger a new era of "gamification" at work. It will also further blur the line between personal space and time and our professional lives. But will machine vision feel like a helpful assistant to aid our work more safely and confidently—or is it poised to take humans out of the loop entirely? This book's second section will pose important questions that come with larger-scale systems around transparency, bias, and equity.

What if robots made dinner—
and did the dishes?

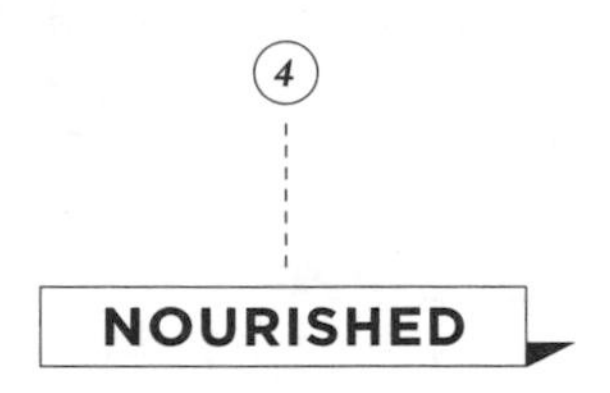

About automated cooking, grocery shopping, growing food, and cleaning up

4.0 Eating with your eyes

shareable meal playlists | *kitchen-cleaning robots*

Our kitchens contain the most expensive tools and technology-enabled "furniture" in the home, and demand a lot of our time and attention every day. There is also a frustrating asymmetry to these efforts: the planning, shopping, slicing and dicing, cooking, and serving take hours, only for our culinary creations to meet their mushy and forgotten end in minutes (at least in my house). Yes, on many days, many parts of the process bring pleasure. Preparing a meal is a multisensory and often social experience that reverberates with tradition, learning, and nostalgia. An ideal future would accentuate the desirable facets of cookery while streamlining the drudgery, like dishwashing. Fortunately, SuperSight presents an opportunity for that wish to come true.

Cooking began simmering toward an automated future back in the mid-twentieth century. Easy-to-use features such as "just add water" and magic-seeming microwave ovens freed housewives from spending so much time in the kitchen and promised more time spent at the dinner table with the family—or in front of the "boob tube" with newfangled TV dinners.

This was not the utopian, egalitarian domestic future that many felt like they were promised. But SuperSight might be able to help us coordinate a healthier relationship to food, without going to automated science-fiction extremes like the dinner pill. Computer vision and robotics can offer selective assistance with a process that consumes more money and time than nearly anything else we do in the home. Consider all the ways SuperSight will help with dinner:

TV dinners swept American living rooms in the 1950s, but overautomated, homogenized, and depersonalized food. They sacrificed taste at the altar of broadcasters' ad-funded business model.

1. **Meal planning:** Deciding what to eat will be inspired by a photo you saw on Instagram or one you took at a restaurant. Computer vision will scan the image, look up a recipe, and file ballpark calories. Or you'll look to the dinners of people you know for inspiration. One of Spotify's most interesting features is being able to listen to the songs your friends are listening to, turning your playlist into a socially guided choice. There are a thousand things you could be perfectly happy eating tonight; why not make a decision based on what your sister made yesterday to great aplomb?

2. **Shopping for/growing ingredients:** Instead of relying on the grocery store for fresh produce, you'll grow more at home. Computer vision–enabled cameras will tend to your indoor plant wall and fertilize the garden on your rooftop or outside your window, or (if you're lucky) in your actual yard. For produce you can't grow at home, robot shoppers roaming grocery stores will be equipped with full-spectrum vision that lets them predict ripeness and select food that will arrive at optimal juiciness—no more hard avocados.

3. **Prep:** Augmented vision will guide you through the complicated dance of sequence and timing to orchestrate a meal with multiple dishes for

multiple people, all with different allergies and dietary constraints. These same systems will help you avoid mistakes (such as burning things or cutting your finger rather than the carrots).

4. **Plating:** Plating will become high art—certainly even more shareable. You'll have access to more reference designs, projected templates, and the ability to auto-share what you've made to your dining playlist.

5. **Dining:** Because of the volume of information and history associated with food, the eating experience might shift further from being utility-only to a social and learning ritual. SuperSight will project the origin of the food, historical associations, family stories, nutritional information, and suggested conversational topics on the mashed potatoes.

6. **Cleaning:** Instead of taking time away from your family and friends, let the computer vision robots installed on the ceiling of your kitchen do the dirty work of cleaning up while you enjoy your meal, safely out of the way. There will be less food waste, too, as they efficiently package up leftovers for tomorrow's lunch and store remaining ingredients for use in future meals.

SuperSight culinary assistance will also feed our biological cravings for novelty and diversity in food. Just as our tastes in music radically expand with increased access and recommender services, so will our eating patterns, and this combination of variety craving, meal planning, predictive shopping, and local growing will intersect in interesting ways with meal kit subscriptions and restaurant delivery services.

Over the past few years, having all the ingredients for your meal delivered to your door for you to cook has become nearly as popular as just ordering delivery. But one of the huge problems with meal kits like Plated, Blue Apron, and Sun Basket is packaging waste. Without a camera monitoring your cupboards, these services have to send you every ingredient required for their recipes. The assumption they operate on is that you have just moved in and your cupboards are bare—why else would they package and send individual portions of two tablespoons of soy sauce? SuperSight will help monitor pantry supplies with cameras on cupboard doors and inside your fridge, then send this information to meal kit services. Computer vision systems will recognize that you already have thyme, spices, and an egg—no need to send a wasteful extra package—and will aggregate ingredients for the next month to deliver all the brown rice they predict you will use, including for those trendy salmon poke bowls. Your kitchen cam will also track your at-home growing wall, where snow peas are just popping up and a few juicy peppers are ripening. Meal kit services will be able to factor in this home-grown bounty so upcoming meals for the family are based on when your produce peaks. Hello, summer squash, heirloom tomato, and basil salad—all they need to do is send you some sweet corn and burrata the week those tomatoes reach peak ripeness.

SuperSight will also help us solve another sustainability problem in the kitchen: food waste. Instead of you trying to remember if your partner picked up more butter the day before so you don't buy some, too, your fridge will tell you—and even automatically restock when needed. Cameras will track when your milk will sour, and add salad to the meal calendar before the kale wilts. Plus, algorithms could bring people together for communal meals based on what's in their crisper drawers and what needs to be used in the next couple of days. You can't make dinner with a potato and an eggplant—but there's a great meal to be made if you combine them with what's in the fridges of people nearby. The excuse to share a meal with others is just a bonus.

The first few chapters have focused on what SuperSight will let us see through glasses and phone screens. In this chapter, we'll dive deeper into how embedded cameras in the kitchen will make all of this possible. How do cameras in the food supply chain see into fruit to predict ripeness? How does a mechanical hand know how to pick up a wine glass as it empties the dishwasher? How does your trash can recognize a banana peel?

First, though, let me introduce you to some robot chefs.

4.1 Burger-flipping perfection

robot chefs without hairnets

One afternoon I was in Pasadena, California, hankering for a burger. I strolled past a Caliburger, and ordered the Double: two all-beef patties topped with

American cheese, garden-fresh plump red tomatoes, and crisp iceberg lettuce. As I started to sink my teeth into that wonderful steaming first bite (why *is* the first bite always the best?), I took in the scene behind the counter. Instead of humans at the grill, I spied something that looked like it should be welding steel or riveting car panels on a Tesla assembly line. One of its cool metal arms casually flipped patties while another hand pulled sweet potato waffle fries out of boiling oil.

The robot, named Flippy, uses computer vision to determine just when and where to place its spatula to scrape up the meat—or plant-based alternative—and place it precisely on the bun. It also performs a job no human wants to do: scraping burnt burger residue off the grill.

As I chewed, I wondered: Does Flippy represent positive progress for foodservice, and for humanity, or is this blatant automation of burger-craft a step too far? I appreciated that no hair nets or hand washing was required, yet my burger was still likely cleaner than the average fast-food shop's would have been . . . and that, since Flippy can see the full UV and infrared light spectrum to determine the heat of the meat, it delivers perfect medium-rare burgers all day long with a smaller chance of food poisoning.

Curious about Flippy's job performance, I asked its (human) manager. "He's a little quicker than a human worker, doesn't mess up as much," and scrapes the grill "better than any employee." Flippy doesn't talk back or complain, doesn't require healthcare packages or paid time off, and never sneezes on the burgers. Flippy also makes the kitchen safer for the remaining human workers—no more grill scorches or oil burns from the deep fryer.

Flippy represents the future of all restaurants, not just burger joints. All over Asia, robotic chefs and servers are beginning to move from gimmick to new normal. Alana Semuels writes of one such Japanese restaurant in a feature for the *Atlantic*:

> The "head chef," incongruously named Andrew, specializes in okonomiyaki, a Japanese pancake. Using his two long arms, he stirs batter in a metal bowl, then pours it onto a hot grill. While he waits for the batter to cook, he talks cheerily in Japanese about how much he enjoys his job. His robot colleagues, meanwhile, fry donuts, layer soft-serve ice cream into cones, and mix drinks. One made me a gin and tonic.

In my home city, Spyce is a restaurant that uses robots to make a variety of delicious meals such as "the Mesa": "Quinoa. Pumpkin mole. Black beans. Roasted broccolini. Lime-dressed red cabbage. Cotija cheese. Corn poblano salsa. Silk chili pepitas. And chorizo-spiced portobello." Michelin-starred French chef Daniel Boulud helped create the menu, the National Sanitation Foundation gave the restaurant's cleanliness its check of approval, and Spyce says its meals are cooked in three minutes or less.

In Paris, a three-armed robot called Pazzi will make you a pizza at EKIM in just under five minutes. And speaking of pizzas, the Seattle-based startup Picnic has developed a machine that can make three hundred in an hour. Just plug in the order and stock the pantry bins with shredded cheese and hunks of pepperoni, and the computer vision–enabled system will determine the size of the pizza and begin laying out suitable toppings.

That computer vision enablement is key. As anyone who has tried to find the biggest avocados for a three-for-one deal can tell you, the shape and size of most produce vary hugely. And the shape and size of whatever you serve vary greatly, too. You can't just program an algorithm to dice onions or "rotate salmon 180 degrees when browned"—the robot needs to be able to "see" what it's attempting to prepare, so it knows whether your sourdough needs another five minutes in the oven or the cake batter needs another quarter cup of milk.

In the years ahead, economic forces and consumer desire for spectacle and theater in restaurants and bars will spawn a cornucopia of food-focused robots far more advanced than Flippy. They will resemble super-coordinated teppanyaki chefs, hurling bowls of rice through the air and cracking eggs simultaneously with all eight of their arms. You'll stroll into your neighborhood bistro to see long-limbed robot bartenders dazzling patrons with their dexterous flair, pouring drinks like the best human mixologists do today.

Robot chefs won't serve up just burgers, but also complex haute cuisine, cooking with a precision that humans can't equal—and no cocky attitude from the chef, either. These robots will know that when you swap in gluten-free almond flour for a celiac guest, you need another half-teaspoon of baking powder to help your muffins rise, and other subtleties of cooking chemistry.

Computer vision is taking on the dining experience, too. Caliburger has ordering kiosks that scan your face, log you in to its loyalty program, and pull up your preferences for quicker ordering. The restaurant even allows you to pay for your order via facial recognition. In Asia, robot greeters and servers are becoming increasingly common (though their novelty still currently outweighs their efficiency). A *Wall-E*-themed food hall, where wheeled robots bring you your noodles via a magnetic track, opened in Hefei, China, all the way back in 2014. Since then, the technology has advanced: Bear Robotics has tested a hospitality robot in a South Korea Pizza Hut that can guide its way through the restaurant to deliver your meal, then bus your table. And in Shanghai, there is an *entirely* human-free KFC.

Because these robots are connected to the internet and know you as an individual consumer, they'll eventually prepare food that conforms exactly to your tastes and dietary requirements. Do you have a peanut allergy the kitchen should be aware of? Did your doctor recently prescribe a low-sodium diet? Are you part of the 5% to 15% of the population who simply hate cilantro? (It's a genetic thing. Really.) The human and nonhuman restaurant staff will be able to see this from your profile—which could eventually be plugged into your medical records—and sprinkle the precise amount of sea salt your body needs.

And then there's the delivery side of things. Now that on-demand food-ordering apps have proliferated, many restaurants' main source of income is coming from outside their brick-and-mortar locations. (A new term, *ghost kitchen*, refers to centralized food-prep kitchens for online orders that don't actually host diners.) But instead of a high school kid delivering your pizza in his parents' Mazda or your chow mein being wheeled to you by bike, robots will take over that, too. Many companies, from Starship Technologies to Amazon, are experimenting with sidewalk delivery robots that could bring your meal right to your door.

About 4 million Americans work in the fast-food industry (though fewer as I write this in the pandemic summer of 2020). That's a lot of jobs to be automated away by literal iron chefs. And it could happen sooner than we think: the CEO of Yum! Brands, which manages chains such as KFC, Pizza Hut, and Taco Bell, says that robots could start making a dent in fast-food labor by the mid-2020s. And a 2017 McKinsey report estimated that more than half of tasks performed by restaurant employees could be automated by existing technologies—and that was a few years ago.

But you only need consider the growing number of tattooed baristas pulling ristrettos in artisanal coffee shops to know that the future of hospitality will be only selectively automated—that human skill and personal interaction will still be valued. Computer vision–enabled robots will automate certain aspects of certain jobs, just as dishwashers and mixing machines did for a previous generation of cooks, opening up time and space for more customization and craft.

I asked food writer and cookbook author Amy Traverso about the most satisfying part of cooking: "There is so much multisensory sensuality in cooking! I would never want to give up moments like drizzling glaze over hot cinnamon buns." Campbell's Soup can't compete with the spiritual satisfaction we get from browning meat over a hot fire or preparing a stew from scratch. We are, in a deeper, subjective sense, *nourished* by these moments. "Automation might be welcome for repetitive tasks like making bruschetta appetizers for 20 guests, the holiday cookie production line, or canning," Amy said, "but I'd like to see a robot chef fold a stuffed tortellini or apply decorative piping with sugar paste on a wedding cake!"

With more culinary coaching from SuperSight, people will have more guidance and confidence to cook for personal expression rather than just utility. And yes, in the process, our kitchens may get a little messy, but fortunately, that's something else robots with SuperSight can help with.

4.2 Another pair of hands in the kitchen

perceptual appliances

As we start to reap the benefits of more mechanical chefs in restaurants, the inevitable technology trickle-down will bring these culinary robots home, too.

In the 1920s, the modernist architect Le Corbusier famously wrote *Une maison est une machine-à-habiter*: "A house is a machine for living in." His vision will turn from philosophical to literal as kitchens are remodeled around computers that can see, touch, and even taste. Robotic arms installed somewhere near the sink, stove, or fridge will keep watch over the kitchen and possess the dexterity to assemble and cook most of your meals. To avoid breaking the good crockery

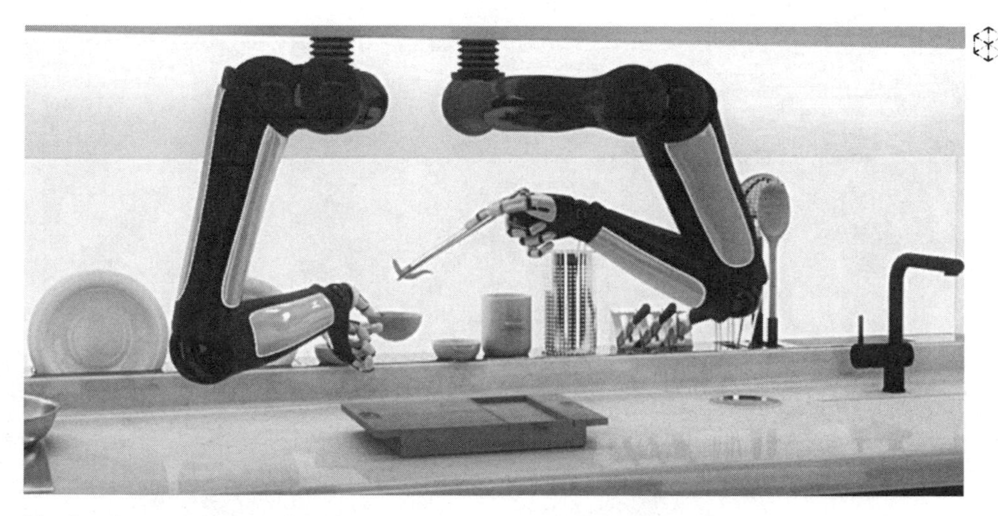

The food prep robot "Moley" has two very humanoid hands trained to mimic the exact motions of a human chef.

or accidentally mistaking the paprika for cayenne pepper, though, they will have to be enabled with computer vision.

There are already some kitchen robots out there *without* SuperSight. Moley Robotics unveiled what it claims is the world's first robot chef at the Hannover Messe Robotics Fair in Germany in 2015. What is so uncanny about "Moley" is the human-like movements of its disembodied arms: it was trained on footage of real cooks in kitchens, so it is actually just puppeting the movements of human chefs. This approach is only good for short demos at trade fairs, since, without vision and haptics (the sense of touch), the utility of this approach is brittle. As soon as it accidentally bumps the salt to the right, it won't be able to find it again. I hope Moley will be given the eyes it needs.

It's not just robotic arms and helpers that will see smarter; our existing home appliances will, too. I want my oven to understand what I'm cooking. I want it to look *into* the molten chocolate flan so the insides remain runny while the crust gets a bit crisp. My microwave should know that I'm about to boil over the oatmeal or explode the egg and save me the disappointment and cleanup. My vent should peer down at the pots and turn itself on and off to vacuum up smoke from my seared salmon. My blender should blend the smoothie just enough to chop the bigger chunks but not into a too-smooth paste. And my toaster . . . well, duh.

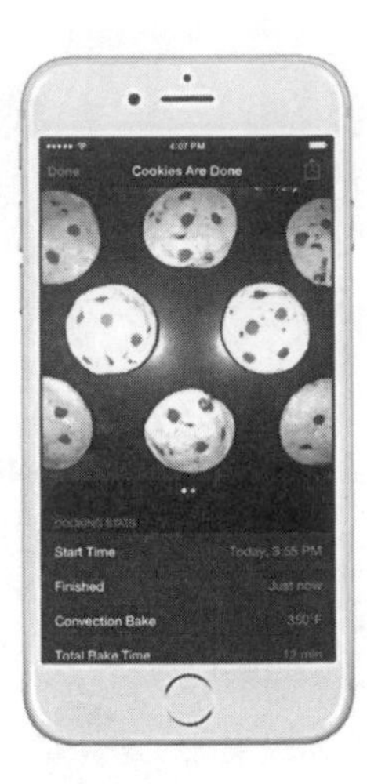

June, the first oven to host a camera inside to watch your cookies.

Soon, expect all kitchen-appliance makers—not just premium brands like Miele, Bosch, and Gaggen-*whatever*—to launch SuperSight-enabled hardware products and cloud-based software subscription services to match. A data center in the sky will process real-time images of burning English muffins and learn how best to respond. Should your toaster whisper through your Alexa to signal that your bagel is turning into a crouton? Or pulse your oven's overhead light? Or just turn the device off?

One of these products that is already on the market is the June oven.

Q: How do you get people to spend $2,000 for a $200 toaster oven?

A: Add an internet-connected camera.

I love to hate this oven. I admire it because a simple internet-connected camera adds a lot of clear value to any oven; once you realize the usefulness of *not* burning your kale or cookies, how could you imagine living without a camera in every kitchen appliance? But as more SuperSight-enabled products enter the market, they won't carry the same price premium. A $5 camera and Wi-Fi chip won't justify an $1,800 markup for much longer.

The way that we read recipes will change because of SuperSight, too. Google, Amazon, and Facebook have already released countertop devices with cookbook apps that check off ingredients, set timers, and auto-advance the recipe so your screen doesn't get smeared with buttery fingerprints. The next generation won't use touch screens; instead, recipes will be projected across countertops and work surfaces, starting with a checklist of ingredients. Recipe-specific portions will be beamed into cups and bowls for measuring (*fill to here*). Instructions will appear where we need them, for example on cutting boards (*chop this thin*). Timers will appear over what we are making or on the oven door, and templates for plating arrangements will enhance your presentation.

The food-experience feedback loop will continue at the dining table. Your table will have a top-down camera that detects the look on your partner's face at first bite. If she looks delighted, your meal-planning algorithm knows to save that recipe and what to tee up for tomorrow. If her face puckers from too much zest in the key lime pie, your kitchen's smart system will suggest zest-adjustments to other dessert recipes.

People make two mistakes when thinking about domestic robots. One is that they assume they must be mobile, rolling around the home. They don't need to follow you around like Rosie from *The Jetsons* or those teleconferencing iPads on sticks. After all, your sink doesn't suddenly change locations every night. The robots don't need to move—only to stay out of your way.

The second, related mistake is imagining that robots and humans enjoy cooking alongside each other. If you make robots strong enough to do anything useful, it's very challenging to keep them from hurting you; they run the risk of chopping

BOSCH produced a cooking concept video to promote their LightDrive AR technology.

you instead of the garlic. This is especially true if you give them sharp tools or anything to manipulate that's heavier than a sock puppet. The best solution is to adhere to that old advice about what to do when there are too many cooks in the kitchen, and have robots do their work independently, safely, and at their own pace.

After your smart kitchen helps you cook your meal, why not delegate cleaning to the bots, too? Rather than rolling, the optimal cleaning robot might be a fixture in the ceiling, where it otherwise masquerades as a swank chandelier. It would unfold from the sky to fill the dishwasher and disinfect counters, return clean dishes, pots, and pans to their proper homes, and maybe even water the plants.

Since you don't want children or pets in the way when a mechanical arm is flailing around, most of this work should be done while people are away from the kitchen. Robotic kitchen arms will use SuperSight to sense that no one is in the room, and a deep cleaning is needed aprés fondue. And since there's a certain pleasure in seeing work being done on your behalf, it will time its actions to finish just as you're coming down for breakfast in the morning.

Concept rendering envisioning a ceiling-mounted robot that fills and empties the dishwasher, cleans your kitchen, and tends your growing wall from multiple anchor points in the room.

Unstructured environments like a human kitchen represent the next frontier for researchers across the field of robotics. Every home kitchen is different. Think of how many cupboards you have to open in a new friend's home before you find the coffee mugs. Kitchen environments are also particularly problematic because robots will need to not smoosh delicate heirloom tomatoes or crack eggs unless it's for your omelet. Any robot that interacts with the physical world faces an inherent uncertainty in how objects will react to touch; even sober humans are known to break wine glass stems.

Not only must a SuperSight system only be able to recognize the object it needs and determine how to interact with it; it must also know where it belongs. You'd be frustrated if your silverware were in a different place every time you came home. The best cleaning crews photograph your home in an ideal state so they can return things to the right place after your wild party; a cleaning robot will do the same, though it might also suggest kitchen storage optimizations based on what you use most frequently.

How much would you pay to never have to clear the table, do dishes, or empty the dishwasher again? Does $5 per meal, or $15 times 365 days times 10 years sound about right? $54,750 is a large upfront cost for most households, but a leasing or pay-by-consumables model (think razors + blades, toothbrush + bristles, Kindle + books, etc.) might decrease the pain of paying by spreading that cost out. Maybe a Procter & Gamble cleaning robot only uses their particular brand of expensive soap, and the company subsidizes the capital costs in exchange for a longer-term and more profitable consumables lock-in. New partnership opportunities exist here for consumables companies to partner with durable goods makers. The Whirlpool SuperSight robot might prefer Brawny and Tide, with subscriptions to these products rolled into your payment for the tidy home service.

Are there any dystopian downsides to kitchen cleaning robotics? If so, it's hard to see them. We'll likely host dinner more often if we don't have to clean up after, and if it's easier to accommodate allergies, taste preferences, and predicted portion sizes. And with more time to spend on community building and conversation rather than cooking, these dinner party experiences will allow for more time indulging in delights beyond food, like education, stories, music, meditation, and more.

But it's hard to anticipate the consequences of new domestic action tools. After the invention of washing machines and vacuum cleaners in the 1940s, futurists predicted an increase in leisure time activities like reading and Parcheesi. Instead, we spend the same amount of time cleaning, because our sanitary standards have skyrocketed. TV dinners and microwaves in the 1970s heralded a world where we wouldn't really need a kitchen at all: just give everyone a fridge and microwave, and a large trash can near the TV tray, and they'll be satisfied and satiated. More than anyone imagined, the kitchen has become the center of gravity of the home, where we eat 90 percent of our meals and gather with friends.

I've argued here that SuperSight will guide and teach us to be more productive and expressive when making food, and that a robot's place in kitchen automation is the undesirable jobs, primarily cleaning. But robotics have another important role to play in food futures, one that takes more patience and diligence than most humans are willing or able to muster: growing and harvesting food in the context of home.

4.3 Farming from home

growing meal kits | *infinitely patient eyes*

I grew up with a garden and experienced the pleasure that comes from eating your own lettuce and Swiss chard all summer, fresh tomato and basil sandwiches in August, and squash and pumpkin soup in the early fall. Planting and harvesting on any scale gives people a sense of pride, an understanding of the toil behind store-bought ingredients, and the freshest-tasting produce you can get.

At MIT, I met a cook and gardener passionate about bringing this experience to every home: Jenny Boutin, CEO of SproutsIO, which makes a home-based aeroponic planter that uses computer vision to maintain optimal growing conditions. The camera in the box has only one job: to take an hourly photo of your lettuce leaves, or whatever you happen to be growing. The device uploads these photos to the cloud, where algorithms compare them with ideal growing photos taken from other planters. Based on this data, the device adjusts its lighting intensity, light color, nutrient mix, and other variables to maintain ideal growing conditions.

"Cultivating food at home consistently and reliably is now possible with technology that optimizes the growing conditions for each plant with the help of AI."

—Jenny Boutin, CEO, SproutsIO

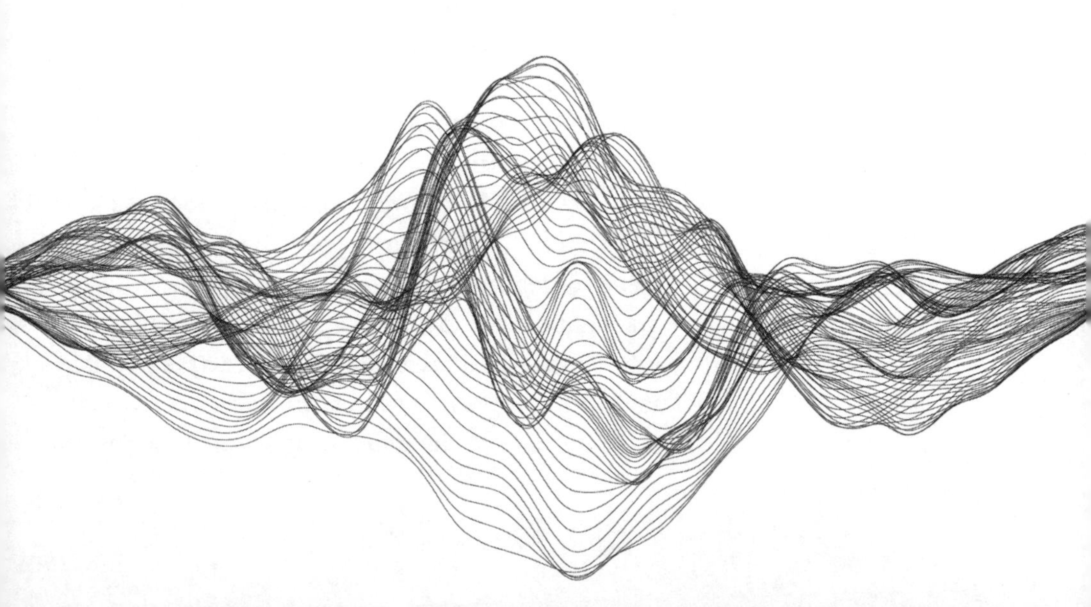

SproutsIO, a planter that spends all day tending your vegetables.

Thanks to systems like SproutsIO, many of the vegetables and herbs we consume can be grown locally inside our home, rather than being driven around the country or flown in. This will transform agriculture as well as the supermarket and restaurant industries by giving us access to what we need, and when. And as more people want to grow their food closer to home—while also being able to travel away from it without coming back to a mass of wilted celery—we will continue to see more opportunities for delegating oversight.

Grow Computer is another company that is using ever-patient AI eyes to care for your vegetables. The idea first came to New Yorker Dan Nelson when he was planning his honeymoon. He'd spent some years previous really getting into home gardening, and had developed quite a nice setup at his home in Park Slope. He'd lovingly nurtured many vegetables from seed—but what were his plant babies going to do for the seventeen days he was on vacation?

Dan started tinkering with a bunch of existing systems—including a failed homemade gravity irrigation platform—but couldn't find one that seemed to work remotely and reliably. He wanted to be able to see his plants from a distance. He wanted to know what the moisture levels were like in the soil. He wanted an idea of the humidity in the air. And he wanted to be able to adapt his automatic watering cycles in reaction to what that data was showing him. It took him four years to get it right, but now Dan has a growing system others can adapt for all sorts of different needs, from watering and temperature and light control to transplanting and harvesting.

A few companies are trying to take what is possible in automated home systems to farm scale. Iron Ox uses computer vision for hydroponic growing systems in custom-built warehouses, with robots to handle everything from planting seeds to keeping plants healthy to harvesting the crops. The growing environment is finely tuned and controlled so plants never experience the extreme heat and cold that can ruin delicate crops like lettuce. Automated agriculture systems like Iron Ox not only expand what farms can grow, but also *where* farms can thrive. Since robot food-growing warehouses can be erected anywhere, why not co-locate them with population centers or at least grocery distribution hubs to keep produce fresh and reduce transport time, costs, and waste?

SuperSight is also now being used in agricultural settings to perfect not just a few dozen home-grown tomatoes or a warehouse full of robot-grown ones, but hundreds of thousands of fields. One major application is the plant equivalent of personalized medicine. Previously, crop sprayers would need to blitz an entire yield with the same amount of nutrients or pesticide. But now computer vision–enabled cameras on these large-scale spraying vehicles can assess the health of individual plants and dole out the right solution just for them. They can also give farmers a heads-up if they spot signs of disease or bug infestations—and nip them in the bud. This means fewer pesticides, healthier crops, and happier farmers.

John Deere is the leader in the field (pun intended). The 180-year-old big-ag company has, over the past decade, been turning itself from a tractor company into an AI enterprise. They are betting big that the future of farming includes a lot of tech. "The revolution taking place with deep learning has opened doors to solving problems that farmers have dreamed about solving for years," John Stone, SVP of John Deere's Intelligent Solutions Group, told *Forbes*. "With computer vision systems and deep neural nets, there's a very exciting future in these technologies in farms." Today these semi-autonomous systems help farmers

See & Spray Select John Deere reduces the use of pesticides by employing SuperSight to recognize the difference between cultivated plants and weeds behind a tractor. Now individual plants can be specifically treated to combat herbicide resistance.

safely navigate vulnerable rows of tightly planted produce and drive autonomous sprayers. This assistance frees up farmers to tend to other parts of the business, because they're spending less time sitting behind the wheel plowing a perfect line to the end of each row.

This is important because the future of farming needs saving; the world is running out of farmers. Plagued by climate change–induced extreme weather conditions and shrinking margins for selling their wares, fewer kids are taking over their families' farms; the average age of a US farmer has risen to fifty-eight years. SuperSight might be the breakthrough technology that, by improving the efficiency of how our food is grown, makes it possible to keep supplying grains, corn, and fresh produce to the billions who need it. And unsurprisingly, SuperSight is poised to transform the way we shop for that food in stores, too.

4.4 When shopping feels like shoplifting

sentient supermarkets *subway station shopping*

Did you know that 3% of all employed people in the US stand at a cash register? In high school, I worked as a cashier in a busy bagel shop, and it was a challenge even for my teenage body: constant standing, risk of cuts, wet hands rewashed many times an hour. Not to mention the nonstop stream of demanding customers, each with their oh-so-particular needs: "A little more sprouts on my veg salt bagel—no, not that many—and could you re-cut the salmon lox a little thicker and add three more capers . . . but hurry, I'm late for class!" Consumers don't like checkouts, either. Long lines are frustrating, the service is rushed, and you're always glaring at a person in the twelve-items-or-less aisle who *definitely* has fifteen items in their cart.

Many parts of these grocery-scanning jobs could be quickly automated. Ringing up orders, counting cash, serving change, and waiting for people to swipe a card rarely leads to meaningful human interaction, anyway. And what a waste of space: think about the number of square feet in a grocery store reserved for expensive cash registers, conveyor belts, and bagging zones.

SuperSight offers an entirely new paradigm that will forever change the grocery store: human- and money-less checkouts. Who wouldn't want to simply browse items, pick up what you want, and walk out of the store, no interaction or credit-card swiping required?

We saw, in chapter one, an example of how SuperSight can be used to remove the friction of paying from the consumer experience: by using facial recognition for micropayments. But SuperSight is also accurate enough to see and record exactly what you pick up while shopping—or just point at with your phone. As an example, Tesco is the number-two grocery retailer in Korea. Instead of building more physical locations to fuel further growth, or bringing more people into their existing stores, they chose to bring their stores to the people. They installed large, lifelike billboards in subway stations that are identical to real stores in scale, but paper-thin. Scan any item with your phone and it lands in your online cart and is delivered to your home, sometimes faster than the train can get *you* there.

In 2016, some of the most talented computer vision PhDs at my social shopping company, Ditto, were hired away to lead a team that commercialized putting SuperSight into the supermarket: the appropriately named Amazon Go stores.

When you enter a Go store for the first time, a nice human at the entrance instructs you to download an Amazon Go app, associate a credit card (buzz kill), and tap in. Once inside, it looks like a normal grocery store (a bit of a

Tesco's employee-free supermarket poster in a subway station. Scan the items you want delivered while you wait for the train.

disappointment—couldn't they do anything more interesting than shelves and coolers?). The idea is that you pick up the items that you want, like a sandwich, an overpriced smoothie, and granola bar, the same way you would at a normal supermarket. Then comes the fun and frictionless part: there is *no* checkout. Simply walk out of the store! A couple of seconds later, Amazon sends your phone a push message with little photos of what you took, and the total with which it dinged your credit card.

I had a chance to visit a Go store in San Francisco in 2019, and I wanted to see if I could beat the computer vision technology. C'mon—I had quick reflexes and a big bag. Could I shoplift?

Once inside, I looked up to see what I was up against. The ceiling was painted black to kind-of disguise a network of cameras and cables: not just a few, but literally thousands of cameras, directed at different parts of each shelf. Each was trained to track people, hand postures, and the presence or absence of items on the shelf—so the AI knows that, if a granola bar was missing, it must've been shopper X.

To test the limits of its skills, I picked up things and replaced them. I picked up things and moved them to other parts of the store. As fast as I could, I snatched a granola bar on a lower shelf with my jacket covering my hand. Sure, the AI cameras could see me, but how fast could they read a gesture? I stashed

An Amazon Go store, where there are hundreds of cameras in the ceiling trained on every object.

about five items in various pockets and made a beeline for the exit. I thought I surely must have gotten something past the cameras. What match was a computer system for sleight of hand? They had just launched the store a few weeks ago; I figured the system must have had some bugs or at least optimization problems to work out.

Then my phone beeped. Damn. The cost of five items had just been debited using my credit card, accurately and perfectly.

Today, Amazon Go stores are spreading to more cities, and offering more fruit and veggies, baked goods, meat, and seafood. The way shoppers handle these, like squeezing the avocados one by one or smelling the melons, puts more demands on the computer vision systems. Dilip Kumar, Amazon's VP of physical retail and technology, summed it up in a Reuters article after the launch of their London store: "There's a lot more interaction that tends to happen with [fresh] produce than a can of Coke."

⚠ HAZARD: PERVASIVE PERSUASION

Your experience of retail will be highly curated by algorithms that highlight specific items of interest. With diminished reality in the supermarket, you won't even see items you're allergic to.

Since SuperSight can see what you are looking at, in stores and out in the world, shopping will become semiconscious, and promotions will respond to subconscious signals of interest. Just stare at something for a few seconds as you stroll through a store and it's tagged. A quick scowl at something else will also be noted.

Of course, a lot of where we look isn't intentional. We scan our surroundings to take in a scene, and are attracted to vibrant color, high contrast, illumination, and especially motion. If a squirrel darts across your field of view, it's impossible not to attend to this moving stimulus (just ask any dog). But it turns out we can learn a lot about a person's level of interest and intention by measuring how long they look at something. A double-take (like a double-click) is even more of a tell: something was *so* interesting to them that they involuntarily took a second glance at it. In other words, the items that catch our eye—a pair of shoes in a shop window, the person at the other end of the bar, someone else's appetizer choice—are a strong subconscious signal of interest.

Marketers love this kind of tell. How long your gaze dwells on things in your environment forms a continuous heat map that could be sold to businesses in two forms: the personal version ("she's interested in X, based on what hooks her attention") and the anonymized aggregate version ("everyone in this demographic set is fixating recently on Y").

Imagine if retailers had access to the data for every single thing you looked at that dilated your pupils with longing: outfits on mannequins, ridiculously priced scallops, the Apple Watch you picked up but then put down. They could choose to dynamically price a product depending on your interest. You might pick up the shampoo bottle you ignored if you saw the price drop. Or if you keep coming back to that same Fendi bag over the course of the week, maybe the price increases to encourage you to buy it now, before it goes up again. Or, if that bag is truly out of your price range, other brands might have your SuperSight glasses suggest cheaper ones that look similar for purchase.

> To protect our choices as consumers, the ability to easily edit and delete what companies think they know about us needs to become the norm, now more than ever.

Over the next decade, most stores will follow Amazon Go and adopt ambient automated checkout powered by SuperSight, and will be reconfigured to exploit the valuable space at the front of the store that registers used to fill. And the staff that stood at those registers? Hopefully they'll be redeployed to other tasks, like providing higher-touch concierge and guided experiences to those interested. I would love more human interaction in grocery stores, offering more tasting samples and expert advice. Give me a copilot to help me find everything I need for a complicated dinner party, or a dietician to help me select food products to meet my health goals, given my family's food allergies and interest in local produce. After all, without humans to assist us, food shopping would become an even more isolating consumer experience. Better to leave just the picking, packing, and delivery to the robots.

What if education were
spatialized and democratized?

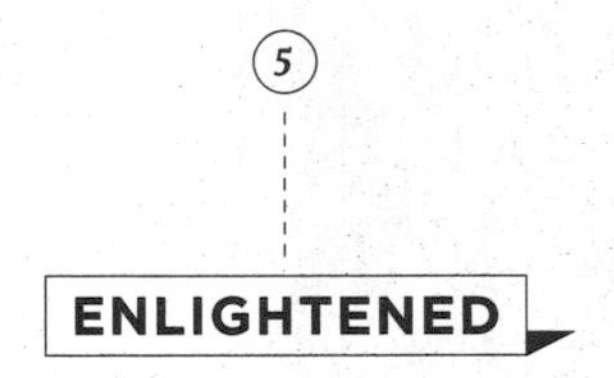

About contextual, continuous, and embodied learning

5.0 Educational tourism sponsored by your airline

auto-captioning *historical reenactment* *microbetting*

Barcelona is an easy choice for a family vacation because SuperSight glasses are part of the package, heavily subsidized by the Spanish travel and tourist board and Iberia Airlines. The airplane steward tells you the lenses will auto-transition to sunglasses when you're outdoors in the Spanish sunshine.

The first feature is immediately helpful: Spanish subtitles. The array of microphones on the front of your glasses picks up on the language around you, then directly projects the translation, visually positioned over the speaker, so you understand your driver's excellent suggestion for a gelato detour. That night, being able to read the ingredients for paella makes it look even more appetizing. And you get snippets of conversations from passersby, talking about secret beaches.

The next morning, you encourage your kids to turn off the auto-translate function so that they can practice their Spanish. (When they get stuck on a word—either spoken or written—they can always call up the dictionary embedded in their glasses to double-check.) Your first stop on the day's itinerary is a Barcelona must-see: Gaudí's architectural masterpiece, Sagrada Família. Using your glasses, you can scrub through time to see the building's progression over the last decade, and the final plan for how it will appear when it's finally finished in twenty years.

Afterward, you decide to go for a walk through the old city, but instead of taking one of those overstuffed walking tours, you download a personalized (super) sight-seeing map, which lets you chase a virtual bull down narrow side streets,

Construction projects will be finished for tourists' eyes, like Gaudí's Sagrada Família cathedral in Barcelona.

learn about the history of key buildings like the Barcelona Pavilion from Le Corbusier himself, and observe reenactments of key moments from Spain's rich history going back to Roman times.

That afternoon, you luck out and get football tickets. Your glasses make the nosebleed seats so much better because you can zoom in on the action, see the names and backstories of each player, and learn about their tactics, like drawing players offside or faking an injury (something that's been elevated almost to an art form by the opposing team from Italy). Since Europe has laxer gambling laws than the US, your glasses help you wager on nearly every aspect of the game, with money or frequent flyer points. *YES, I'll take those odds for winning the penalty kick. Gooaaaaaal!* (You remember just in time to switch on the parental lock in your kids' glasses.)

The experience for both you and your kids is better than any summer school. They use Spanish for most interactions, are immersed in a very different culture (the best way to understand it), and learn about the country's street food, architecture, history . . . and obsession with *fútbol.*

The line between school and the rest of life is about to blur on a massive scale, giving rise to numerous new business models and consumer offerings and extending the potential for online learning brands. Leading scientists and other star communicators will become celebrity educators through future tech-enhanced glasses, and learning will happen everywhere around us, continuously, throughout our lifetimes. Education will be more vital than ever for personal satisfaction and career success, but the desks-in-a-row classroom settings and rigid multi-hour schedules we associate with learning will morph into something more ad hoc, continuous, and atomized. Education will be situated in places like museums, national parks, and city streets, and events like concerts, parades, and protests will gain a rich patina of historical explanation to make them more interesting—and enlightening.

Computer vision will make education for all of us much more playful. Think back to kindergarten: learning was more about making and building and following your curiosity, as teachers facilitated. Soon, your glasses will play the teacher's role. At the art museum, they might guide you through self-portraiture by female artists or early photographers experimenting with silver halide chemistry—whatever you're into. If your gaze dwells on a particular work of art, they'll offer up the artist's motivation and information on what other works or experiences inspired them. At the symphony, your glasses will project the history of instruments, background on the musicians, and a diagram that explains the structure of the piece, pointing out the recapitulation of the main theme. Concerts will also be "annotated," with brief notes on music theory to keep you engaged and enrich your understanding.

You'll be able to have similar learning experiences closer to home. Want to know whether that spider under your child's bed is dangerous? Your glasses will identify the spider's species, and explain that those little pods hanging in the web are just about to burst into thousands of tiny replicas of Mom . . . plus that she just ate the husband for protein. If, after binge-watching *House Hunters* on HGTV, you're inspired to learn more about the architecture and flip-ability of houses in your neighborhood, you can enable the Zillow filter for your glasses on your evening walk around your neighborhood. They'll tell you why the bungalows from the early 1900s are more affordable, while the American Colonials tend to hold their value, as price graphs sweep up next to each house with a sales history and predicted maintenance costs. A neighborhood filter on your glasses might also introduce you to the owners and backstories of the small businesses you pass, using life-size, lifelike holograms.

The trend in education toward affordable access and bite-sized learning is already well underway. Consider the information available online on YouTube, through Khan Academy's videos, in short courses from EdX, Udacity, or MasterClass, and on one-minute "daily dose" podcasts. What is different with SuperSight is that opportunities for augmented learning will be more situated and spatialized. They'll be triggered by and embedded in everyday situations: that evening walk through your neighborhood, making a smoothie, watching a ball game, listening to a street performer—even sitting in the very chair you're reading this book from right now (those chair designers have a lot to say about their process and material choices!). And education that is situated will be more memorable and effective.

5.1 Now you're on the Magic School Bus

simulation-based learning | *episodic memory* | *multimodal encoding*

The night before my brother-in-law's paintball birthday bonanza, I read up on the sport. I learned all about the gear, the CO_2-gun technology, how paint is pelletized and released at different ranges, team strategies, and how to avoid an ambush. I was going to *nail* this party, so Sam wouldn't keep calling me his old academic brother.

The next morning, together with fifteen of Sam's friends, we donned combat fatigues and helmets, loaded up hundreds of paintballs into semiautomatic gear, and split up into teams. They read us the rules about not shooting each other in the head or neck and sent us off deep into the woods to take our starting positions. When it came to the reading, I couldn't have been more prepared. But when the starting horn sounded, I panicked and cowered behind a barrier as the action played out around me. I was scared and overwhelmed. When I finally summoned the courage to stand up and peer over the packing crate, I was nailed in the clavicle with a painful ZWAP within five seconds, and a bright blue splotch of paint marked me as out of the game. I wasn't the courageous hero I'd imagined.

Playing against my brother-in-law's friends was totally unfair, of course; they're all firefighters and other first responders. I had never been trained to think critically when my heart rate was over 80 BPM, whereas they already knew how to perform in physical and mental extremes. Their preparation was dominated by simulations and rehearsals—exactly the approach required to succeed in the high-respiration rate, high-blood-pressure, high-stress, deep-woods sport of paintball.

Embodied cognition is the difference between reading a book about paintball and playing paintball. No matter how many strategies you read, or how many diagrams you absorb, they cannot prepare you for being pelted by primary-colored paint pellets. These are two different ways of learning and knowing: theory and practice.

Traditional schools often fail us because they place too much emphasis on theory—too much Googling about paintball and not enough playing of it. So what if you took the books out of it entirely?

Mixed reality (that blending of virtual and real through SuperSight) is poised to remake how we teach and learn through what psychologists call *grounded cognition*. The theory is that meaning is grounded in, and understood through, our senses and episodic experiences. For example, when you see a photo of a dog, you may recall the feeling of petting fur, the smell of dank breath, the wet warmth of a tongue, or the sound of a whimper or bark. Contrast this seeing with a photo of a dinosaur: unless you've been to a particular theme park on an island off Costa Rica (or have a time machine in your backyard), your knowledge is more theoretical, without the sensory encoding and richness.

When we passively listen to teachers and are tested on our ability to memorize facts, we only use what psychologists call *semantic memory*: facts, ideas, meaning, and concepts. But the most effective educators and training programs understand the difference between memorizing something and *learning* it. Remembering the rules of paintball is one thing; applying them in context is quite another. Modern tools like SuperSight let us boost semantic memory through immersive technology, engaging us in ways that lay down a more durable type of memory: *episodic memory*. For example, instead of learning the year the Declaration of Independence was signed from a printed page, you could use AR glasses to enter a simulation in which America's forefathers are debating states' rights. You're in the room where it happened. The episodic memory of the debate is much more likely to stick with you than a paragraph in a textbook.

There is a strong link between how and where you learn something and the quality of your recall. This is known as the *environmental reinstatement effect*. If you typically sip green tea when studying for a math exam or listen to piano while teaching yourself Mandarin, you'll be more likely to remember that elusive formula or the right verb conjugation in those same settings. The neurological explanation for this is key to understanding the educational potential for SuperSight: memories are automatically encoded with a multisensory track. You can't selectively learn math without also laying down the feeling *warm hands* and the taste and smell *green tea* at the same time. The brain is adapted to absorb and combine visual, aural, and other sensory context cues, and to use any or all of these as hooks for future recall. We remember more effectively when we encounter knowledge in rich, real-life contexts. When we are *doing*—when learning is participatory, multisensory, emotionally engaging—we integrate information from our eyes, our ears, our body position, others who are present, and much more.

In other words, we encode memories in a multimodal parallel stream. Doubtless you have smelled something that triggered a strong visual memory, or heard music that recalled a relationship. Even a repetitive motion like using a tool or swinging in a hammock can flood you with past emotions. Cognition and memory are embodied.

So now you know why field trips are so much more compelling than textbooks for learning environmental science, history, astronomy . . . nearly any subject. And SuperSight will make hands-on, field-based learning richer and more personal. With smart glasses, old trees and new flowers can tell you their stories and introduce the botanists who first cultivated them. Rocks, fossils, and ancient lichen can play time-lapse views of their history. A tiny beetle can blow up to the size of a house to show you their pincers.

Students will engage with these experiences in novel ways, too—interacting with, capturing, annotating, and sharing them. Their notes, sketches, and detailed audio memos will all be stamped with the time, location, and their emotion. Later, they can search by where they were, who was nearby, and what excited them the most—or where they zoned out ("Siri, show me all of the things I wasn't particularly interested in but that I need to know for the exam—and make them more interesting, please"). SuperSight and AI are getting better at summarizing a story with a time-lapse of the most salient information, and examining critical moments in detail by creating super-slow-motion replays of phenomena otherwise too fast to see. This personalized field-trip media will be shared on social media feeds and liked, reshared, and remixed as avidly as TikTok videos are today. Because of this digital fluidity, self-expression will be a larger part of

"If you have a magic school bus, it doesn't make sense to stay in the classroom."

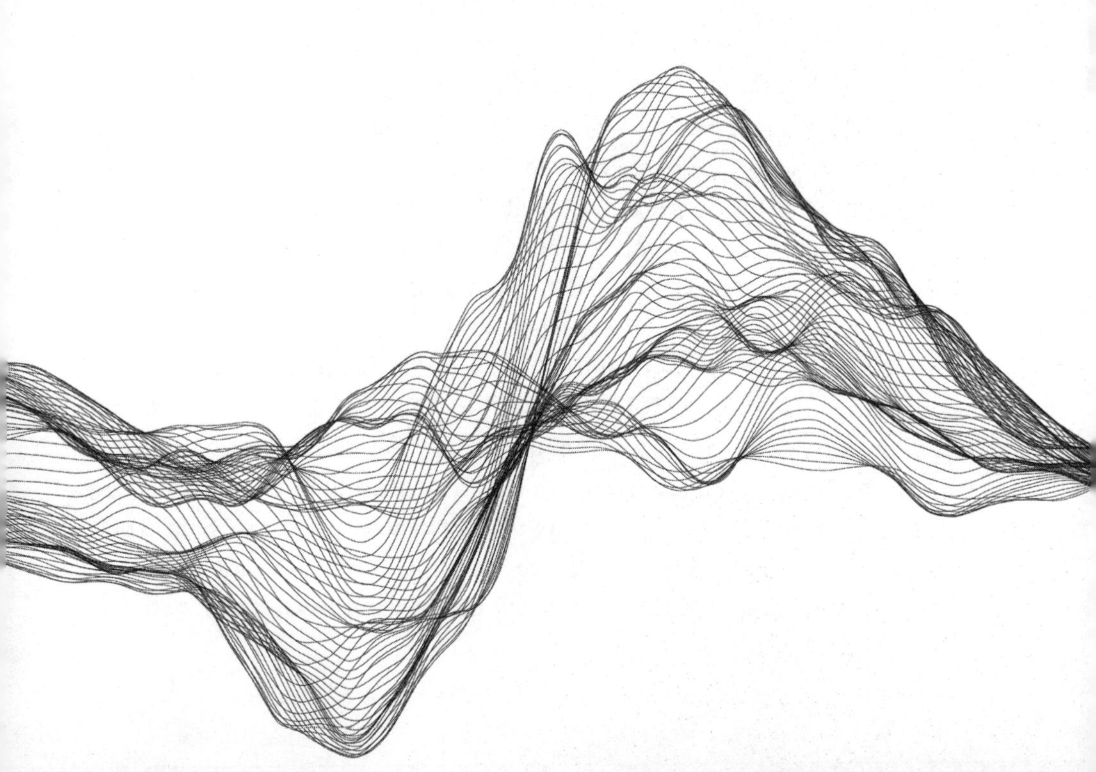

educational experiences—and educational experiences will be an effective promotional channel, too.

I visited a Harvard lab to see how VR is used to teach molecular chemistry. Each student wore a VR headset, and was rotating their head to look up, then at the floor, then behind them to grab and arm-wrestle with the bonding angles of proteins. Students loved it; they truly get a sense of the remarkable 3D structure of complex molecules when they are as large as a truck and they try picking them up to dock them together. But I was struck by how much these VR simulations tend to be individual. Students were shouting at each other—"Whoa, how do you get a hold of the caffeine molecule?"—but no one had a shared view, including the teacher.

The future of classroom education should be social and collaborative—where teachers, fellow students, and the work are all visible. The best labs and learning spaces, like the Graduate School of Design at Harvard and the MIT Media Lab, are intentionally designed to show activity, process, and work. They have literal glass walls and long views into, across, and through labs, workshops, and maker spaces. You don't hear everything, but you do get a sense of visual omniscience. This transparency and porosity helps good teachers gauge when they need to guide, and inspires students by letting them see each other's work. Everyone feeds on the energy of common action. Augmented reality makes that kind of shared real-world view possible in a way virtual reality does not.

Today, to teach fourteen-year-olds anatomy, we kill scores of frogs to fill each classroom with their rigid, formaldehyde-laden bodies. Part of the reason dissection is so popular is that it brings frog anatomy to life in a way pictures in a book can't. But iPad apps like Froggopedia are even better anatomy teachers, because you can perform a dissection while the virtual frog is still jumping around. Peel away the skin and see the rear leg muscles in action; understand why some bones need to be thicker; see the moment-by-moment changes in the heart, lungs, and other organs. In fact, why would you choose to slice open a frog at all, when you could slice open and step into something from, say, the Jurassic era instead? With AR, your class could learn anatomical structure and Darwinian specialization using a full-scale T-rex that sprints toward you on the soccer field. Look at those massive rear haunches, suitable for chasing down prey—those powerful jaws that crunch bones and tear flesh. Now that's evolutionary optimization!

Dissection of an AR frog. In this step, the skin is transparent to reveal the muscular anatomy.

If you have a magic school bus, it doesn't make sense to stay in the classroom. Just ask Ms. Frizzle. By the second page in any Magic School Bus book, she has the kids on the bus and off to some fantastic field trip through the human bloodstream, the solar system, or the city's water pipes. SuperSight will give every student their own magic school bus, and in the process, turn the whole world into a spatial learning experience.

What your smart glasses will have that the magic school bus doesn't is personalization. The supplementary knowledge whispered in your ears about what you're seeing will be personalized to advance your understanding without overwhelming you. This automatic leveling is a big deal. Because they are so close to your skin, brain, and eyes, glasses know a lot about whether you are engaged and learning: galvanic skin response looks at the conductance of the skin, which is a signal for excitement; brainwave sensors in the arms near your temples read if you are in generative flow modes (beta waves, at 14–40 Hz, are associated with reasoning, and gamma waves, at 40 Hz or more, with bursts of insight); and eye trackers measure your pupillary saccades to identify your precise subject of interest. An algorithm that combines these could dynamically adjust a lesson's content depth and pacing, or switch approaches when needed.

These augmented layers will do much more than show static labels. They will also make time and space elastic to reveal a world previously invisible. One day in my high school physics class, Mr. Brunschweig switched off the lights and started a reel-to-reel film projector to show us a film called *Powers of Ten* by Charles and Ray Eames. The movie takes you on a remarkable journey through scale: every ten seconds, it zooms another order of magnitude, out to the edge of the Milky Way, and into the nucleus of a carbon atom (that was as far as science could see in 1977).

SuperSight, like *Powers of Ten*, allows us to see the world not only at different size scales, but at time scales, as well. An architecture student with smart glasses will be able to scrub through a timeline of how a building was constructed with the same two-fingers-extended gesture that DJs use when scrubbing a vinyl record: the building will disappear down to its foundation and then be rebuilt in time-lapse. What history student wouldn't love to bring up a "time knob" that lets them experience incredible historical moments: moving speeches by John Kennedy or Martin Luther King Jr., as they were delivered on the Mall in Washington; Thomas Edison or Nikola Tesla's lab at the moments of invention; a castle siege or a Revolutionary War battle. Being able to get *in the simulation* will turn everyone into a history buff.

Voyages across time or size scales make the world feel more legible, more interesting, and appropriately multi-layered with history and accessible stories. Thanks to the holographic experiences SuperSight enables, anything with a digital model can be cut away or blown apart into its component pieces, or seen through entirely or selectively. One of my favorite examples of *literally* embodied augmentation is an anatomy-teaching shirt made by a company in London called Curiscope. When you point your phone or iPad at a fellow student wearing the T-shirt, you suddenly see into the chest cavity. There, a beating heart pumps as blood rushes in and then is expelled out, on its way to 3D expanding lungs for

The classic Eames film *Powers of Ten* illustrates how we might zoom in to the subatomic scale and telescope out to galaxies to learn with SuperSight.

A student wears a Curiscope AR T-shirt that exposes their anatomy and skeletal structure in AR.

oxygenation. Every part of the circulatory system is sized and positioned approximately, and in the right plane, for the body wearing the shirt; your screen shows half kid and half skeleton.

Imagine being able to X-ray *anything*. What might we learn by walking around our built environment with X-ray glasses? It's a Richard Scarry cut-away fantasy: slice through the layers of concrete, asphalt, and rock in the city to see the subway network, water and sewage, gas, and trains flowing below you. See into walls to view wiring, plumbing, insulation, and critters. Car engines, airplane wings, elevator pulleys—with SuperSight, the way things work is structurally revealed.

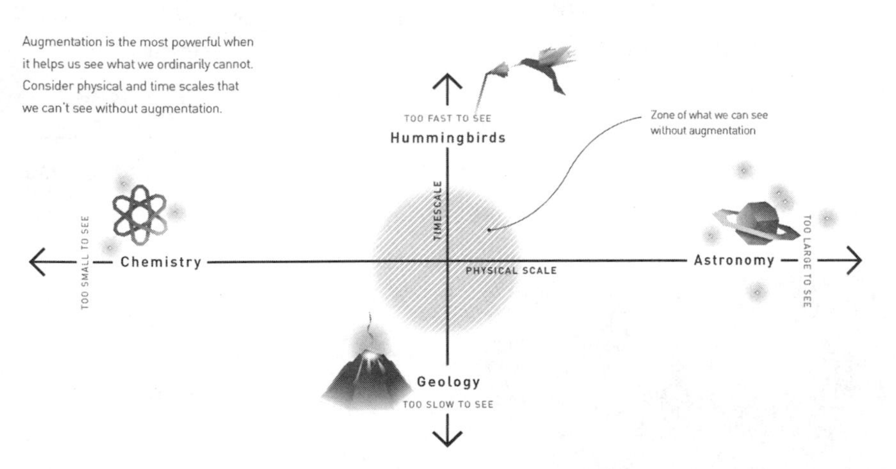

When teaching with AR, consider the elastic potential to slow down and speed up time to explain phenomena.

The best AR experiences will take us into phenomena at scales of time and space we wouldn't ordinarily have access to, from small (molecular scale) to large (astronomical scale), and fast (hummingbird-wing scale) to slow (sea-level-rise scale). Nearly everything is more impressive under a microscope or from a God's-eye view, in super-slow-mo or compressed in time. SuperSight will allow us to feel like we're in *The Matrix,* where we can slow down to *bullet time,* and to time-travel at will—whether we're in the classroom, or outside of it.

5.2 Turning the city into a time machine

spatial history lessons *armchair archaeology*

If you get the chance, I recommend visiting the One World Trade Center tower in New York. Constructed on the site of the Twin Towers destroyed in the 9/11 attacks, its design is so well considered and fortified that it took more than a decade to build. As you ascend to the observatory, on the 100th through 102nd floors, you're in for a great surprise. Each of the four walls of the elevator is augmented with a data projector that shows a 360-degree time-lapse view of the post-colonial history of Manhattan.

You enter the elevator below ground, so you start off looking at the cold, dark, dripping bedrock of the island. Then the doors close and you feel the elevator rise and break free of the ground. As you begin to fly upward, you see the East River surrounded by a pastoral landscape with a few small cabins, fenced-off plots with livestock, and a church. A small indicator shows that you're in the first decades of Manhattan's founding by white settlers in the early 1500s. Time progresses quickly as you rise over a growing settlement; a network of roads begins to wind

The Freedom Tower elevator uses projected augmentation to animate the history of Manhattan over three hundred years as you ascend the building.

north. As you ascend farther into the sky, the years rapidly scroll by, buildings get taller, and streets denser. The experience is breathtaking as it unfolds all around you. As with the ascent of the Great Glass Elevator in *Charlie and the Chocolate Factory,* the view is more panoramic the higher you go; the expansion of the city also seems to accelerate. When the elevator slows and the doors finally open into the observatory, you exit with a hundred questions about city planning and what Manhattan might look like in another fifty years—a totally different mind state. Certainly it's been the best elevator ride of your life.

Let's unpack why this elevator trip is transformative. First, there's a correspondence between the place and the augmentation. There's a perfect alignment here between where you are, what you feel, and what you see (something my electronics friends would call an *impedance match*). Because you can look in any direction, the experience feels incredibly immersive and true to life. The orientations correspond. And you have the natural haptic pressure on your body as it rises in space. Another reason why this elevator is so spectacular is that you don't start off with a view—you start off in a claustrophobic granite cave, and then get to experience a sense of relief and elation that architect Frank Lloyd Wright called *compress and release* as you break free of the ground and rocket upward.

The One World Trade elevator experience points to a future for augmented education where idle moments can morph into powerful learning opportunities. Whether we're taking public transportation, standing on an escalator, or waiting in line, augmentation can teach, inform, and inspire.

A Plexiglas sign superimposes this line drawing over a triumphal arch, the last of the remains of an entire Roman city in Austria—one piece of inspiration for how SuperSight can reanimate history.

SuperSight holds enormous promise for learning about history and culture in its native context. Seeing the remains of long-ago societies leads us to envision the past and wonder what our lives might have been like, had we been born in a different time and place. SuperSight can manifest simulations to help us explore these questions, starting with the *power of place* and building up from whatever stones remain.

Imagine the first-person scenarios that could unfold as you tour the crumbling, dank castles of the British Isles, the Olmec heads in Mexico, Masada in Israel, the Great Sphinx, Stonehenge, Easter Island . . . Millions of tourists visit these archaeological sites every year, but the short leaflets and earnest docents can't fully bring them back to life. SuperSight offers incredible potential to vividly render and immerse visitors in the culture and pivotal moments in history—not only as bystanders but as participants. Old castles could be reanimated with a cast of thousands for you to learn about medieval festivals, social hierarchy, armament, and more.

For those who can't visit historical sites, there are many VR applications that allow you to pretend that you're Indiana Jones. One of the biggest is Open Heritage, a collaboration between CyArk and Google Arts and Culture. Using 3D imaging, they replicate dozens of historical sites for anyone to view in VR, including the ruins of the Mayan city of Chichen Itza in Mexico and the earthquake-ravaged Bagan temples in Myanmar. John Ristevski, CyArk's founder and CEO, was born in Mosul, Iraq, and was inspired to start this project after the Taliban destroyed 1,500-year-old structures in Afghanistan in the early 2000s. He is now working to record many other existing sites before they're damaged or worse due to climate change or war.

VR also allows students and armchair archaeologists alike to experience sites that have already been destroyed. Through a variety of apps and museum experiences, you can explore Bronze Age houses, Qing Dynasty porcelain factories, and Herod's Temple from the first century AD. In 2015, UNESCO and the Institute of Digital Archaeology used 3D scanning and printing technologies to replicate Syria's Arch of Palmyra, a third-century Roman structure destroyed by ISIS. These physical sites and monuments may have been eradicated from the physical world, but not the virtual one.

SuperSight is also assisting real archaeologists like Stuart Eve, one of the directors of the current excavations of the battlefield of Waterloo. In his blog *Dead Man's Eyes*, he describes using Apple's ARKit to digitize much of the data from the Waterloo excavations and layer it both upon the site they're working in and back in their studios. Archaeologists only get a few days or weeks on site each year, so this life-size digital replica is a game changer, allowing the work to continue back home.

SuperSight's ability to see into the past isn't just about structure and appearance of places; it's about interaction and dialogue with the people that lived there and learning what drove them. A few years ago, at Plymouth Plantation outside Boston, I met Charlotte, a thirty-four-year-old woman from 1660 who had just endured a monthlong Atlantic crossing and was struggling to harvest enough food for the harsh New England winter. She wore an uncomfortable woolen anorak and used words like *chirping-merry*, the opposite of *mulligrubs* (down in the dumps); *swill-belly* (a drinker); and *roast meat clothes* (Sunday best). She sleeps, she said, on uncomfortable mattresses made of straw in a dark room with only a tiny smoky glass window. Charlotte was an actor, but soon you'll be able to find virtual versions of this type of historical simulacrum, with their Revolutionary shoes and scratchy wool dresses, along the Freedom Trail in Boston engaging with the tourist hordes. They'll tell stories of revolutionaries throwing tea overboard and give impassioned speeches inside the old meetinghouse. These historical ghosts, reanimated and true, are going to be amazing. The modern buildings will fade away, and based on the year you choose, people and events will be brought back to life.

SuperSight has the ability to turn real places into history and science museums, but what might it add to the learning experience for actual museums, libraries, historical societies, and other places dedicated to preservation?

5.3 Making museums magical

augmented exhibitions | *deepfakes* | *portable dinosaurs*

Museums have always been about a certain type of time travel. Art museums beam you back in time to appreciate revolutions in artistic rendering, from

Dadaism to pointillism to abstract expressionism. Natural history museums compress time to help us appreciate natural selection. Science museums highlight changes to the physical world over time, for example in evolution or to Earth's constantly shifting continental plates. And these time machines would be made all the more believable through the drama and explanatory power of AR. Using the incredible exhibits they produce as the backdrop, augmented layers will add essential context and teleport us in new ways.

We've talked about SuperSight giving you X-ray vision to see through and into bodies. In the Smithsonian Museum, visitors use augmented vision to do just the opposite: add muscle and scales to the bones of prehistoric fish. "Fleshing out" is a term used in the real world for adding context to something. In this case, it's not a metaphor—we're literally adding flesh to help us see the world in a new way.

With SuperSight, we'll apply flesh to ancient dinosaur bones, and see the reincarnated pterodactyl shatter the glass of its exhibit case in the paleontological wing to exit the building. Whether enabling you to wander around the museum's halls from the comfort of your couch, or offering in-person headset displays before you enter an exhibit, many major museums are experimenting with virtual add-ons. What are the best models for taking advantage of these incredibly storied buildings? How might we better bring the past into the present, by merging the digital and the real?

A visitor uses augmented reality to return flesh and scales to a fish in the Hall of Bones at the Smithsonian.

Maybe you should ask Salvador Dalí.

Dalí wasn't exactly modest. And he'll tell you so, too—to your face. Despite the fact that the famously neurotic Spanish painter passed away in 1989, the Dalí Museum in St. Petersburg, Florida, has created an experience that makes you feel like you're chatting with the Surrealist.

Berkeley students suddenly appear to have professional ballet skills using a computer vision technique called generative adversarial networks.

To bring Dalí back to life, the exhibit designers used a burgeoning AI technology to create what's called a deepfake. This portmanteau of "deep learning" and "fake" describes taking pieces of media from multiple sources and synthesizing an image or video with an uncanny potential to deceive its viewer—like bringing back a historical figure to teach you about art. Talk about your *surreal* experiences!

Researchers first showed the potential of the generative adversarial networks that make deepfakes possible by applying the stripes of a zebra or the spots of a cheetah to a video of a galloping horse. The result was a highly realistic but entirely confusing new film. The instant mash-up works on humans, too: in her PhD project "Everybody Dance Now," Caroline Chan, a sleep-deprived Berkeley grad student, seems to become Bruno Mars as he dances to "That's What I Like," then pirouettes with the pose and skill of a professional ballerina. Her machine-learning PhD is paying off, at least in terms of YouTube fame!

As part of the Dalí Museum exhibit, you can take a selfie with the late Surrealist.

Of course, there are other, less humanitarian uses for deepfakes. Yes, these synthesized films can teach history through reincarnating Genghis Khan, but they also put professional models out of work. When your deepfake can dance like Bruno Mars, why pay a real actor who isn't as talented? Most troubling are the fake news videos of politicians convincingly saying things they never uttered, and fake sex tapes where porn stars' bodies are merged with celebrities' heads. In response, the DEEPFAKES Accountability Act was introduced in the House of Representatives in 2019, and several states also have introduced legislation to limit their use.

Still, the educational potential is undeniably exciting. Inside the Dalí museum, his deepfake stands next to you, pondering his Surrealist canvases of melting clocks. He's up to chat about anything, as long as it pertains to his life and art. He'll talk about painting technique; his wife, Gala; or his brother, who died before he was born: "I wish to prove to myself I am not the brother, but the living one," he says in a meta-reference to his unreal nature. "The only difference between a madman and myself is that I am not mad."

The *pièce de resistance* is when you exit through the gift shop. He thanks you for visiting and beckons you closer to take a selfie together. After you've snapped a picture of you smiling together, he oohs and aahs over the photo, and offers to text it to you. Then, he gives you his signature sign-off: "Kiss you, bye-bye!"

The most stunning thing about this experience is that you feel like you're interacting with the artist himself. It looks like Dalí, sounds like Dalí, and even seems to move like Dalí. "Some people have cried," says Hank Hine, executive director at the Dalí Museum, as reported in *Smithsonian*. "Just the fact that someone has been resurrected, it's pretty amazing. It has this spiritual impact. If you can see Dalí come alive, then why not believe in resurrection, eternity and your own immortality—and the immortality of those you love. It's very uplifting."

Imagine if museums provided this kind of experience for every historical figure, not just artists. (Well, maybe not Genghis Khan—and definitely not Charles Manson.)

Museum visitors will soon be able to use interactive SuperSight tools to extend their museum experiences beyond the institution's walls, too. Here's a quick scenario: Fourth-grade science teacher Sruthi discovers an immersive AR experience at her local natural history museum. She takes her class for a visit, where her students cut and paste 3D animated models from the museum into their digital notebooks. Then, working together back at school, the kids copy those augmented reality dinosaurs onto the school playground and buildings, setting them loose to wander where their fellow students can experience them, too.

"AR technology will enable rich immersive learning experiences in our homes, neighborhoods, playgrounds, parks, and more, literally bringing experiences once reserved for certain places, for certain people, to everyone everywhere."

—Michael Soileau, vice president, Comcast

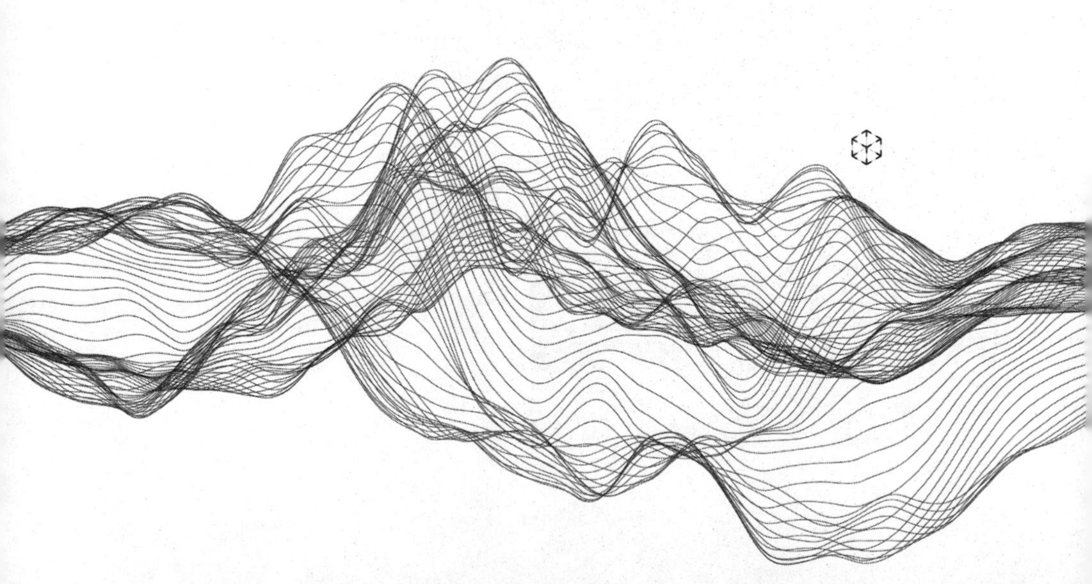

Concept sketches of a learning app where you can copy dinosaurs from the museum and paste them into the schoolyard.

Educational use is just the beginning for this type of tool. Forget postcards at the gift shop; a simple framing gesture will virtually cut and paste anything you see, in 3D, into your living room or wherever else you choose. This will be especially fun for massive sculptures like Claes Oldenburg's *Spoonbridge and Cherry* at the Walker Museum in Minneapolis. Imagine dropping that in your front yard or on the grass outside your dorm window.

There are so many ways to use SuperSight to enhance museums. If your museum has bones, add some flesh and animation. If your museum has characters, animate them and let them become guides. If your museum has multidimensional artifacts, allow visitors to not only take 2D photos, but also to cut

With SuperSight, massive sculptures will be scanned and projected into other contexts—with or without permission.

and paste 3D versions into their notebooks and share 3D objects on their social media streams—and add a cheery Dalí to just about anything.

"Kiss you, bye-bye!"

5.4 Enchanted objects that help you learn

educational toys *projected light fields*

Interacting with the subject you're learning about in the ways just described creates the kind of episodic memories that make new information more memorable. But an even better way of laying down episodic memory is to give people the experience of invention. Let them make mistakes rather than try to learn from errors others made in the past. Let them learn by doing.

This is the primary learning approach used at the MIT Media Lab. Its mantra of "demo or die" means that students are rewarded not for talking about an idea or making a PowerPoint proclaiming its coolness, but instead by going down to the shop to create a physical object as a demo and programming in the functionality they imagine the object will have. Through this process they learn the limitations of the materials—and often whether their idea was viable at all. They're much more likely to remember why an idea, interaction, or design didn't work if something goes wrong with their project, versus having an instructor like me lecturing them about its potential problems. They learn firsthand what's challenging and what works more easily than anticipated. These usually end up being throwaway projects, but the result is a tremendous amount of learning by doing.

One semester at the Media Lab, I was teaching a class in which the assignment was to create wearable information—jewelry, clothing, or other adornments that dynamically change to express some piece of online information. One student team had the idea to make a jacket that responded physically to social media engagement. They started by using a little pneumatic pump that added air to the jacket; if someone gave you a like on Facebook, the jacket would puff up a bit to mimic your inflated ego. This was a fun concept, but they quickly realized a couple of things as they moved to a prototype. First, pneumatic pumps are kind of heavy, require a lot of power, make noise, and aren't exactly machine washable. And second, social media likes are a slippery currency, easily faked when expressed in internet-connected apparel. If super-puffy jackets became a visual shorthand for popularity, someone might just buy a marshmallow-like jacket *without* an internet connection. The project built cross-disciplinary skills in hardware hacking, coding social media APIs, wearables, and the non-STEM topics of culture, garment design, and identity politics. As an educator, I'd say that's a pretty rich mix.

At Continuum, we use *lighthouse projects* to help companies envision the medium- or far-horizon future. What products and services might they offer in five to ten years? What will the brand stand for in that future? The metaphor of a lighthouse sprang from the old adage that large businesses are like large ships—it's very hard to get them to turn, but once they do, they can exert incredible momentum in the market. So, rather than trying to get them to turn little by little, incrementally innovating a client's existing product lines, you envision five or seven years out to create a wholly formed vision for the company's future. This lighthouse articulates a vision: a safe harbor toward which they can steer. With its destination clarified, the company often realizes it needs to start funding whole new divisions or acquire new technologies to better compete in this future world.

One such lighthouse project was for the toy company Fisher-Price. For decades they've sold plastic toys, but we wanted to help the company imagine a future in which toys exploited innovations like 3D printing, the Internet of Things, responsive materials, projection mapping, and more. Through a series of workshops, we brainstormed scenarios that would rethink key moments in children's and parents' days, then infuse them with new products and interactions.

At Continuum, we designed a set of augmented future-of-play experiences as a lighthouse project for Fisher-Price. Here, a child's growth triggers projected photographs of their earlier growth milestones—a spatialized, tangible entryway into your photo album.

We weren't designing any specific product that Fisher-Price might market, but instead creating a pastiche of options that would inspire product lines they might invent in the coming years.

In the first scene of the future of learning-toys scenario we created, a child arranges thick, felt Colorforms-like elements to compose a friendly owl, which triggers a 3D knitting machine to "print" her bespoke plush pet. In the next scene, a projected-light playroom has a *Where the Wild Things Are* woodland scene displayed on the walls. When the child is eating, projected light draws their attention to certain foods, explains the origins of where they came from, and highlights essential vitamins. At bedtime, each turn of a book's pages triggers new ceiling projections and ambient sounds that complement the story. The future of learning, we felt, should be generative, participatory, and embedded into how we live, fusing traditional materials and familiar gestures with augmented experiences.

SuperSight tools and capabilities have incredible potential to improve the ways we learn and democratize access to that learning. Governments and school districts could subsidize the hardware and AR cloud of content experiences for kids and teachers; departments of education could fund promotion-free virtual layers for anyone under eighteen. Given the educational power of augmented reality for lifelong learning, and the need for workforce training across many industries, progressive governments would do well to subsidize smart glasses and access to content for adults who can't afford them, too. With such strong leadership, SuperSight glasses and AR-based learning will help kids and adults alike not just learn, but really *understand* the world around them. Education will become more participatory, more collaborative, more scenario based, more theatrical, more narrative—and a lot less boring.

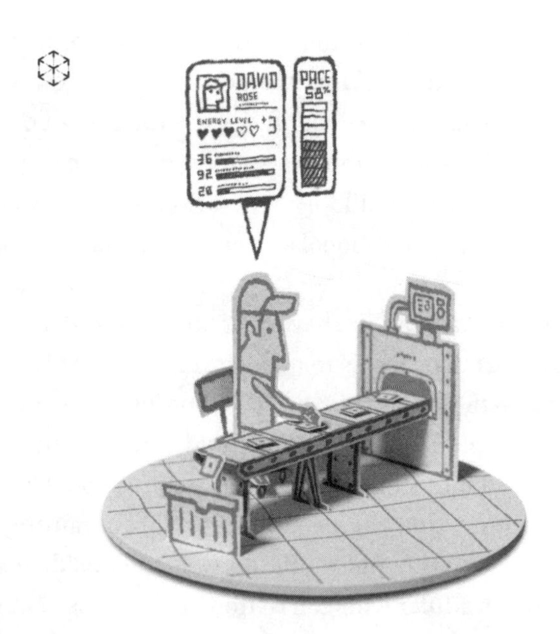

What if your job felt
like a video game?

About gamification and spatialized telepresence at work

6.0 Follow the fox

gamification wayfinding

Go here, now go here, now go here. So many jobs require wayfinding.

In September 2019, I was invited out to Google's Silicon Valley HQ to discuss gesture-based interactions for a smart glasses project. While I made my way through the massive Google campus the morning of my meeting, even Google Maps got confused. I soon realized why. As the algorithms and I tried to navigate to the right complex, we passed a long line of spinning cement mixer trucks queuing up for the next pour. The campus was changing. A new Google-city was rising, funded by your attention, to house the next iteration of the search giant, which has aspirations of moving beyond the tired formats of banner ads and the business model of promoted search results.

When I arrived at my meeting, I learned that Google had hired a massive team of hardware and software engineers and interaction designers, and was investing billions to develop the next-generation AR headset aimed not at gamers but—surprisingly—at workers.

Rather than trying to design for some abstract set of possible applications—one of Microsoft's mistakes with the launch of HoloLens—Google had shrewdly decided to design and prototype their AR glasses for a specific job. They chose the one that was happening all around the Google campus and required huge amounts of 3D spatialized information: construction. And since Google was the rich and ready client, they could easily partner with the construction company leading the vast, audacious expansion to their campus in their bid to reinvent and revolutionize how we build cities with SuperSight.

Specificity and constraints energize designers. Focus on one persona with one issue and you're more likely to solve a meaningful problem. For the Google AR team, that specificity was the problem of job-site navigation. After interviewing and observing many construction professionals, the team learned that getting to the right place at the right time on job sites is an everyday challenge. It's more important than safety, communication, documentation, or a dozen other potential AR applications. Findability is frustrating and inefficient today: workers need to find materials that have just been delivered, and must get to the right place in a building that keeps morphing around them. And construction requires a specific sequence: walls need rough structure, then plumbing, electrical, inspection, insulation, drywall, painting and finishing, and finally fixtures. Coordination and timing for these jobs is highly choreographed, and without wayfinding, everything backs up: teams are forced to wait, schedules fail, profitability falls, and material is wasted at near-criminal scale.

Industrial SuperSight glasses will be a huge help in coordinating this intricate place-based dance across workers and trades. That is exactly what Google is prototyping today, as a case study project for the larger world of work. With AR help, the plumber knows where to find material dropped off for them, and where to prioritize their work this morning so they don't delay the electricians, the inspectors don't slow down the insulators, and the drywall crew doesn't hold up the painting schedule.

Construction may be the most complicated situation for AR navigation, because of the numerous trades at work, the material supply chain for each, and the ever-changing routing environment. But navigation is an essential part of even most white-collar jobs. You have to get to the right conference room at the right time to present or to meet a recruit for an interview. In certain jobs, navigation *is* the task. Delivery people need to know the best route to a set of destinations given traffic; doctors and nurses need to quickly navigate complex hospital hallways to reach patients with critical needs; and managers keeping tabs on progress need a sense of who is where in the organization to casually intersect with clients and staff. We all need a dynamic map for work, and AR might offer the best way to stay on track.

Since the invention of cartography, navigation's dominant metaphor has been a treasure map: plan view, top down, X marks the spot. But using a map is complicated. First you must build a mental model of the space based on the map's scale (does an inch equal ten meters or one hundred?), then figure out your place and cardinal heading to depart in the right direction and at the right pace to arrive at your destination on time. Maps of large airports often include notices like "20 minutes to get to Gate X from here" due to ambiguity of scale and the many kilometers of airport shopping ahead of you. It's so much easier to rely on someone else to do all this mental math and just follow them. SuperSight enables this less cognitively taxing, more delightful future of navigation, inspired by video games and maybe Tinkerbell from *Peter Pan*.

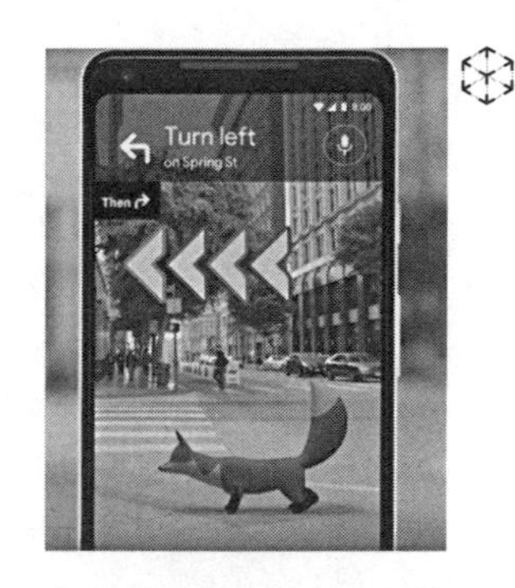

Google navigation will soon be gamified to find conference rooms and people.

Soon, in fact, we'll be able to just *follow the fox*. The future of navigation was presented at the Google Developer Conference keynote in 2018, where vice president Aparna Chennapragada took the stage to demonstrate a new massive machine learning project. Google was training Google Maps to visually recognize every streetscape in large cities like New York, so you can hold up your camera and use that spatial recognition for a new kind of navigation. Rather than the familiar map with a blue dot for "you are here," you'd see a virtual red fox appear on the street ahead of you. The fox happily trots ahead of you while you walk, and looks around to wait for you if you linger behind. Just keep up and you'll get to your destination on time.

This demo should blow your mind, not only because of the technical feat, or how much game conventions like AI-driven characters and avatars are leaking into life, but because of what it means for the reinvention of maps.

SPECTRUM OF REPRESENTATION

(aka, South Park to Jurassic Park)

When you can project any digital information you want into the world, you have a choice to make about virtual style and realism. Clearly a flat, two-dimensional style will be more visible and more differentiated from the real world.

Following a character versus a blue dot on a 2D map is ten times less cognitive work because there is no abstraction of a place that you must hold in your mind. You don't need a mental model or to transpose a 2D schematic plan onto the actual 3D world in front of you. You trust your guide knows where they are going and simply tail them. If you are navigating through a conference crowd, you'd prefer the guide be tall and distinctive, like how tour-group leaders hold a sign or flag above their heads to help throngs of people make their way around the Louvre. Instead of following a sign, why not race after a giraffe?

Navigation-by-character is just the start of how SuperSight-enabled video game patterns will transform how we work. The gaming-trope takeover will impact everything from performance evaluation to where and how work takes place. SuperSight makes it increasingly possible to be seen, digitally modeled, and measured, and this data will be fed right back to us (and others) via our glasses and other place-based projections like conference room tables. Work is a system that loves hungry feedback loops; SuperSight enables both the inputs (sensing) and the outputs (display), and is adopting the visual language and conventions of the world of gaming to do so.

In this way, the worlds of work and gaming are colliding and combining. Through SuperSight, the forces that enable this mix—digitization, accountability, and "countability"—will accelerate, with ramifications for the work we choose, how we collaborate with others, and how we are compensated.

Let's *follow the fox* through a few of the biggest consequences of this shift, starting with the one most people recognize: gamification.

6.1 Follow the leaderboard

workplace gamification *social incentives* *design principles*

One warm summer evening a few years ago, I was strolling with my family down Commonwealth Avenue in Boston's Back Bay. It was dusk, and we were unsettled and alarmed when a horde of people engulfed us, passing by like a pack of bison. The group of forty or fifty people was too large to be a bachelor party, and moving too fast to be a political protest. It wasn't purely a running race, because I could hear people planning and strategizing together as they jogged or power-walked in zigzagging routes. Moving in close formation, they all had their smartphones out and were gazing intently at their screens every few seconds.

What was going on here? I put the question to a straggling member of the herd. Someone yelled, "It's Mew and Mewtwo!"

Now I understood: they were playing *Pokémon Go*, the addictive scavenger hunt game where you find and try to capture adorable animated Pokémon characters sprinkled around the world. My son had just aged out of an obsessive Pokémon card–collecting phase (thankfully), so I recognized "Mew and Mewtwo" as names of two rare and elusive characters in the Pokémon gaming universe. Apparently, someone had spotted the characters on their Boston map, and these enthusiasts were hunting them down in augmented reality to score major points.

Watching this group of strangers immersed in hot pursuit of a shared goal, I realized just how potent a force gaming can be. In this case, gamification was not only reanimating a public park after dark, it was motivating participants to work together, get quite a bit of exercise, and learn about their city and its history in the process.

Games are good for more than just entertainment. Gamification can be a psychologically effective motivational tool. Game mechanics appeal to the human desires for competition, socializing, mastery, status, self-expression, and closure. Research shows that these techniques stimulate specific neurological structures in our brains with predictable benefits like improved performance and greater satisfaction. And as evidence of the impact of these psychological tools has accumulated, they've been increasingly applied to the domains of marketing and business, from frequent flyer programs and Dunkin' loyalty cards to Airbnb reputation ratings.

Consider how gaming metaphors and tropes—points, levels, badges, streaks—have invaded our social relationships, ecommerce transactions, and media viewing rituals. Not only are we already in the habit of rating and ranking our lived experience (e.g., assigning stars to our Uber driver or last restaurant visit on Yelp) and using others' ratings when making purchase decisions, we're also already hooked on the immediate dopamine rush that comes with receiving five stars back. Gamification has crept into most social interactions and relationships as hearts on Instagram posts, retweets, likes, friend counts, and followers, all indicating some sort of proxy for popularity.

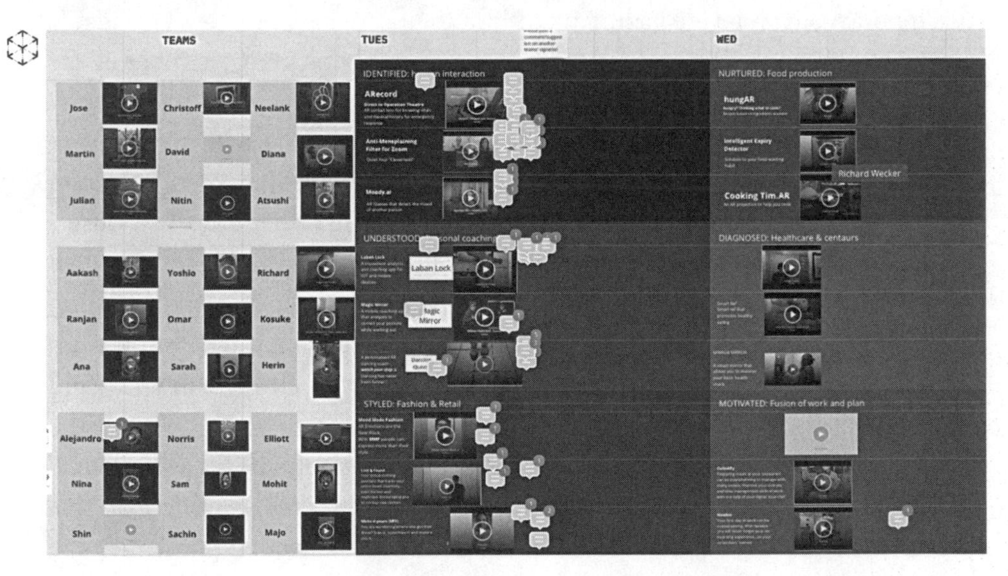

Social-comparison game mechanics used for online teaching. In a recent class of twenty-seven master's students from the Copenhagen Interaction Design Institute, individual and team assignments were posted to a MIRO grid, an infinite canvas that permits everyone to observe everyone else's work in progress.

As social animals, we are driven by pecking orders and social comparisons. When I teach a course, my first step is to set up a system that allows students to see each other's work as they submit it. This amps up their social-comparison brains and drives them more than any inspirational soliloquy I might offer to be more creative and inventive, and to stay up late learning new tools and polishing their work. Instead of me motivating them, they're motivating each other. It's a natural driver of performance, especially when I add comments and stars and encourage students to do the same.

One of the game mechanics that reliably motivates us is leaderboards. When we see ourselves stacked up against others—or even against our own personal best—we are desperate to climb that board. Even for things as mundane as household chore charts, we want to climb over the next schlump on the list. Once SuperSight allows our every action and every move to be captured, logged, and analyzed without any effort, it will send this instinct into overdrive.

The same kinds of point systems that drive the level-up economy of games are now poised to sweep into the workplace, ushered in by a focus on big data and pop-psychological notions from behavioral economics. Gaming's visual vocabulary of stars, badges, and levels will merge with corporate KPIs and ROI measures, with additional accountability coming from SuperSight. Conversation by conversation, meeting by meeting, our job experience will turn into a video game, with our fill-bar progress and "points needed to level up" projected for all to see.

The business approach of *management by objectives* aims to make goals explicit and measurable, even when it means inventing units for things that don't seem naturally countable: presence, confidence, goodwill, engagement. In both games and disciplined organizations, the near-term mission is made clear for the players and measurable by systems, whether that mission is defeating a dungeon boss with fire arrows or winning over your real boss by hitting sales targets. Given this compatibility, gamification has been embraced in business contexts to measure activity, spur motivation, and drive corporate outcomes.

One example of this is Microsoft Outlook's preposterous weekly summation of "focused time," provided as a pie chart to you and your manager. Microsoft surely realizes the hubris and limits of its purview. I, for one, focus in the shower, where Microsoft can't (currently) monitor brainwaves. But as data becomes available from more sources—from smart watches and phone apps, and cameras that digitize analog activities like employees sketching ideas with #2 pencils—the accuracy of that focused time metric will improve.

Digital dashboards, once used to monitor server performance, are now offering people-performance metrics, such as "Sprint Velocity," for agile programming teams, and time spent on email versus meetings. And as sensors reach and monitor more, the layers of managers in the data-dashboard feedback loop are also growing. All this employee data makes managers giddy—these dashboards start to feel like a fun flight simulator, with switches for bonuses and dials for hiring plans. But for those surveilled, much of this data feels beside the point: *BFD that I didn't get the cybersecurity training badge. I was hired for my critical thinking skills and the quality of my ideas, not HR training about email safety.* For gamification to be truly constructive, companies need to measure and incentivize the right things.

In sales, one of the most quantified and competitive job fields, managers use quotas, shared whiteboards, and sales bells to motivate their teams. SuperSight projections will inevitably amplify these systems with the names and faces of new clients floating above salespeople's heads like balloons when they close a deal. Writers at news organizations might display that day's tallied word counts on their backs, or have audience engagement stats emblazoned on their shoulders. Managers could opt in to have projected military-style epaulets showing their P&L numbers for the quarter. Of course, as gamification metrics become more visible in the office, spatially anchored over cubicles and on office doors, and even presented over people's heads at all-hands meetings via SuperSight glasses, they also have the potential to gain pernicious *Scarlet Letter*–like power and bring new levels of anxiety and hyper-competitiveness to the workplace.

The motivational power of gamification is irrefutable: accountability, transparency, and better performance for individuals, teams, and organizations. But researchers studying the "gamification effect" are finding that while some people respond well to the points, streaks, and potential for rewards and recognition that gamification allows, others feel oppressed when managers track performance in detail. For them, accountability triggers feelings of anxiety and fear.

There is also a legitimate concern that gamification may trivialize serious issues, reinforce a cutthroat mindset, and contaminate authentic passion and intrinsic motivation. Will HR's mandatory antidiscrimination training badge make the process of becoming a better and more tolerant person a game to be won instead of a worthy lifelong aspiration? Too much emphasis on extrinsic rewards may pollute intrinsic drive.

The design challenge here is to make meaningful work more entertaining and rewarding, and provide genuinely positive feedback to those who deserve it, *without* using these tools for social control. If we're not careful, what is meant to provide motivation will become the electronic equivalent of a horsewhip.

The key to leveling up our work lives through computer vision–based gamification is to build game elements around what the research says works best:

1. **Focus on achievable carrots, not sticks**. Positive-bias feedback promotes goodwill and loyalty versus resentment.

2. **Include frequent level-ups, or finish lines, and ensure they are meaningful**. People need positive feedback at least every couple of weeks and also definable end dates for certain goals—but these can't be just achievement theater.

3. **Project information into the environment** versus hiding it in an app. One of the powers of AR is to make data unavoidable, which is key if you hope to change a daily behavior as a result of daily information.

4. **Feature the people in the same office or team on social comparisons like leaderboards**, not those in faraway offices who employees don't know. A smaller pool of competitors makes passing someone more motivating.

5. **Present spatial information in prominent places**, like an entrance lobby, to feature larger-than-life people and stories that exemplify the values of the company. Rotate these stories often to spotlight new faces.

I called up a lifelong game designer at Hasbro, Diane Shohet, to ask her about the fusion of work and video games. What design principles should we learn from the experts? "Game designers often include a 'catch-up' factor so that people who lag far behind in a game don't lose faith or motivation," she noted. She used the board game Chutes and Ladders to illustrate this: "Even if you are way behind, there is still a chance to land on a ladder square and climb ahead of the leaders." The lesson here is that games need regular "resets," so employees can start over to rebuild their reputation points, or get a fresh start on meeting quota, with every quarter. Without this reset feature, poor performance becomes locked in like a caste system, and good performance becomes a permanent brand on which someone can glide without further effort. In either case, the motivational benefits of gamification are lost.

6.2 Augmenting offices with ambient information

pre-attentive processing *responsive furniture* *privacy by design*

Leaderboards and other gamification tropes applied in the workplace today tend to be corny, garish, awkwardly applied afterthoughts. How might we motivate and incentivize with feedback that is less conspicuous and more dignified?

Video games offer both a model for providing progressive guidance and hints for beginners, and the algorithms do it only when it's needed so they don't drive off the road. Using the same tactic, job-specific information in the office should be rendered as close in time and location to the task as possible. The design challenge for these augmented experiences at work is to go for subtlety, by presenting ambient information.

The path to subtlety starts with understanding how the eye scans a scene to take in information. While most of us believe that we have panoramic vision, the eye only sees a very small patch of the world at any time in detail; most of our receptor cells are packed into a tiny area at the back of the eye called the fovea. Because the view of the fovea is so narrow, we rely on saccades and head movements to gather information, which is patched together into a cohesive scene deeper in the brain.

Understanding saccades has major ramifications for AR headsets and projections. If you know where someone is looking—which is easy with an eye tracker—you can spend all your computational resources rendering the small foveal view at the highest frame rate possible, and spend fewer resources rendering pixels and detail outside of this zone.

“What useful information might you display in someone’s field of view that provides contextual assistance without distracting them?”

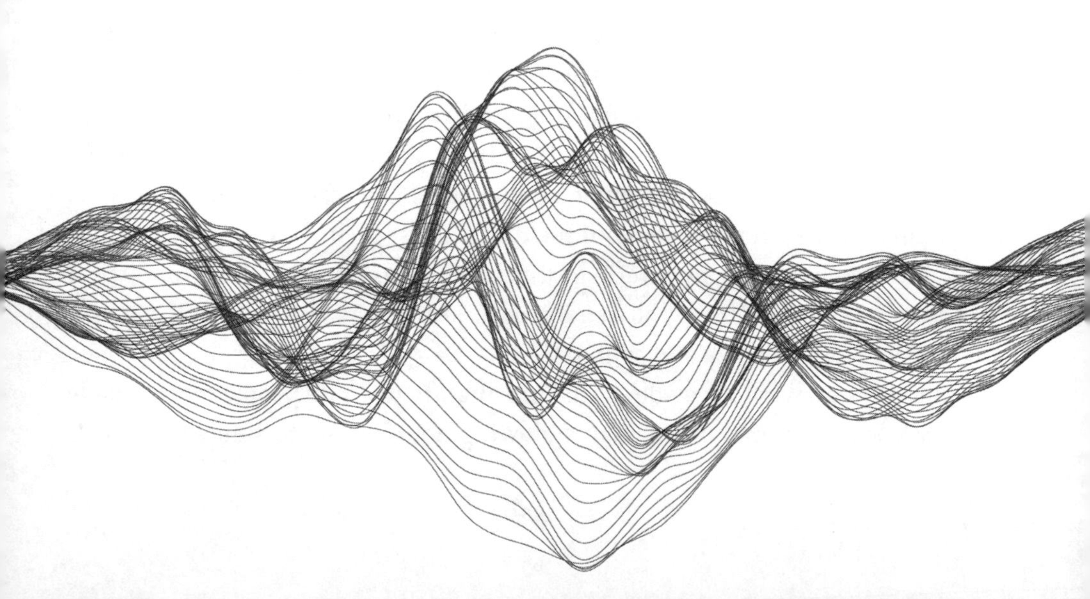

Outside the narrow cone of foveal vision, the eye isn't blind; it's just adapted to process patterns and movement rather than detail. And designing information for this 180-degree panoramic stage presents an incredible opportunity. The world, and every surface in a workplace, becomes a canvas on which to render ambient information.

I first learned about ambient information when taking a cognitive psychology course at Oberlin College, where I worked for a year before graduate school. Between writing software to help people learn Italian and collaborating with the theater department on responsive set projections for modern dance, I was introduced to *pre-attentive processing*, a concept that would change the course of my research and entrepreneurial life.

Cognitive science research into pre-attention shows that there is a set of visual phenomena that our brains process at a very low level, very quickly—in less than 250 milliseconds. And we do this processing in parallel with other tasks, without any impact to our cognitive pool. Numbers and text need our "real" attention, as they take eons (relatively) for our brain to process. Other visual information, though, like color, shape, pattern, angle, and movement, we process pre-attentively, without an incremental cognitive load.

Learning this was a eureka moment for me then, and it presents the essential question for AR designers today: What useful information might you display in someone's field of view that provides contextual assistance without distracting them? If human attention is the scarcest resource, then the best way to respect people's time and mental energy is to make information glanceable.

I realized there was a universe of ambient information displays waiting to be envisioned and brought to life, and spent the next decade designing them at my company Ambient Devices. At Ambient, we created pre-attentive displays like the Ambient Orb, the Ambient Dashboard, and the Ambient Umbrella. Visual indicators on these and other, similar objects tracked hundreds of types of dynamic information throughout the day: stocks, weather, company metrics, steps walked, blood sugar levels, next-meeting countdowns, NextBus arrival times, and hundreds more, all rendered as subtle changes in color, pattern, angle, size, height, speed, and sound.

Our mission was to provide subtle information to guide people making daily decisions with minimal distraction: a glowing door handle to signal someone's sleeping inside, energy-conservation information adjacent to a light switch, a spotlight on your Wellington boots if rain is forecast. If you were building a similar ambient information company today, you wouldn't need to make dedicated objects at all; instead, you could position ambient information with data projection, or use the edges of the glasses frames to position cues in the wearer's peripheral vision.

The applications for ambient information projection at work are vast. You could project an orange ceiling over your desk, for example, when you're in flow mode, to communicate to others that you shouldn't be interrupted. Or paint a green circle on the floor around the coworker you really need to chat with at lunch, to make sure you don't forget.

After Ambient Devices, I was an Academic Fellow at the international architecture firm Gensler, known for their excellence in workplace design. They had just finished designing the Facebook and Pinterest campuses, and were interested in embedding company and team metrics into workplace architecture. Collaborating with their client team at Salesforce, we identified a set of workplace issues where gamification and ambient displays might make a difference. I had just read the book *Quiet* by Susan Cain, who says that in every organization there are introverts and extroverts. You hope the introverts share their ideas, but they are chronically suppressed by the extroverts. Teams and organizations suffer due to this information-sharing asymmetry. Inspired by this problem, I sketched out an augmented conference-room table to help introverts get a little more airtime.

The Balance Table looks like an ordinary piece of furniture, but is actually a real-time ambient feedback device that helps everybody in the meeting see who's been vocal over the last seven minutes. As someone speaks, a constellation of subtle lights appear and accumulate in front of the speaker. Those who chronically dominate what are supposed to be collaborative meetings would be able to glance down, see all of the augmented light, and, assuming they have an adequate amount of emotional intelligence, pause to give others airtime as a result. And if you've been quiet and want to contribute, you're more likely to speak up, even interrupt someone else, because everyone in the room has evidence that someone else has been talking for too long.

The Balance Table is an example of augmenting physical objects in the workplace. But SuperSight glasses, because they'll know what blank spaces are in

Prototyping the Balance Table: wiring up a lattice of LEDs under the table's veneer at technology and design firm Tellart.

The Balance Table augments a conference room table with data to encourage airtime equity.

front of you, will animate many more surfaces in the office with useful information: floors, ceilings, door handles. Your coaching cues might be positioned strategically behind people in conference rooms; your teammates' affective auras might be legible as color to give you more emotional intelligence in conversations, and your software development team might see your coding progress at a glance through projected task cards like Post-its on the wall.

When envisioning the future of augmented reality, people tend to imagine projecting onto static surfaces. But don't forget about the opportunities for objects in motion: a door just as it swings open, the back of the person walking in front of you, objects you hold, or the canvas of your own body. The floor on the way to the conference room could preview the names and faces of clients and the "temperature" of their recent expressions. If a presentation or pitch is in process in a conference room, the door could be covered with yellow and black caution tape, or the wall might have a red "ON AIR" glow.

How you find your way through a labyrinthine office will also change. A vector will show you the directions to your next meeting, its length proportional to how fast you need to move to make it on time. When you reach the conference room early, you can project your key talking points, and the faces and names of those still inbound, on a wall or tabletop, complete with examples of questions each is predicted to ask, to prepare.

As my friend Liz Altman from Motorola recently quipped, "There is a macro opportunity in micro time slots." She meant that these interstitial time windows, like the moments before a call or during elevator rides, bio-breaks, and snack times, are massive opportunities for AR preparation cues or glanceable information layered into the environment.

Ambient information can help us maximize these kinds of moments. But there's another SuperSight application poised to eliminate some of those in-between times, the ones related to transit, altogether.

ANCHORING CONTENT IN THE REAL WORLD

In VR, a software developer can "place" content wherever they want in the fully simulated world inside light-blocking headsets. But with AR, you also see the real world, so smart glasses need to know precisely where to place content in order to complement the natural scene. An AR system needs something on which to *anchor* the content. SuperSight must be used to perceive the world, its lighting, and its objects to convincingly scale, occlude, and place digital elements.

Software developers have a few choices when anchoring and aligning digital content against the real world. They can use explicit visual marks, like a QR code; horizontal or vertical planes; or simply a GPS location. Each approach has different uses and features.

1. **Image anchors** are a good place to start. This is the approach we used for the cover of this book, the chapter head animations, and diagrams. This anchor associates a 2D image found in the real world—a book cover, magazine ad, a printed sign in your office, a billboard, or a building facade—with digital content.

2. **Plane anchors** are useful if you want to "land" content on a horizontal, vertical, or mid-air plane, like a piece of furniture on the floor, art on a wall, or a scene on a tabletop.

Horizontal
For tabletop and horizontally-based experiences.

Vertical
For experiences anchored to a wall or vertical surface.

Image
For experiences attached to images in your environment, like a poster or a book.

Face
For augmenting faces using the TrueDepth camera.

Object
Anchor your experience to a scanned object.

In AR, content can be anchored to horizontal or vertical planes, human faces, specific objects, or specific locations in the world.

3. **Face anchors** are typically used by front-facing cameras to build an AR experience around your face. You see this with social media filters and Apple's Animoji.

4. **Object anchors** detect a 3D physical object in the real world and build the experience around this. This technique might be used to anchor AR content to statues, instruments, factory-floor machines, or other objects with unique shapes.

5. **World anchors** use a combination of sensors to position digital content over outdoor spaces like the sky, fields, lakes, city parks, or mountain ranges. GPS signals localize where you are standing on the globe within a few meters; if you are within range of a 5G mobile phone tower, this standard will provide even more location specificity. Some mapping platforms like Apple, Google, and Microsoft also have enough "street view" data that your camera can recognize precisely where you are in the world given the building patterns or scenes appearing in front of you.

These spatial anchoring techniques are the critical tech that enables us to position and persistently fasten information onto the world. You can leave a virtual Post-it note on the door of a conference room—and know it will still be there when you return the next day.

6.3 Better than being there

telepresence | remote collaboration | gaze vectors

In the movie *Kingsman: The Secret Service,* our hero slips on his telepresence glasses to attend a virtual meeting—one that's much cooler than your average Zoom room. After sitting down at his shiny IRL conference room table, he sees the rest of his agents each teleported into the other seats.

These kinds of telepresence boardroom fantasies have always been curious to me. Why does everyone need to be wearing three-piece suits? (Or are those just virtual, too?) And why enforce a formal, embodied seating arrangement at a table? Are they planning on eating together and need a place to set the main dish, salad plate, soup, and wine glasses? If you're able to teleport, why not meet *at* the topic of interest, like the construction site you're discussing, or the building you're about to use your superpowers to break into—or the top of a volcano or while skydiving, just for fun? And what's with the glowing green translucency? We *know* the other people aren't actually there—do holograms have to forever be glowing green or blue just to cue their "hologramness"? It's been more than forty years since Princess Leia's crackly blue-tinted plea, "Help me, Obi-Wan Kenobi, you're my only hope," projected by the ever-dutiful droid/multi-tool R2-D2. *Star Wars* took place a long time ago in a galaxy far, far away—have we really not progressed any further with the hologram thing? Moreover, the reason we like being with others isn't only to see each other face-to-face; it's a little more subtle than that. We benefit from observing how pairs and groups of others interact with each other, and how people examine and respond to things. What specifically piques their interest and curiosity?

A project called ClearBoard helps solve this *gazing* problem. When its creator, Hiroshi Ishii, first arrived at MIT, he was known for his research on remote collaboration. What might a video-conferencing setup look like, he wondered, where two people are creating and making something *in that moment,* not just communicating an existing set of ideas? This project continues to inspire me, especially after 2020's year of non-stop Zoom culture.

ClearBoard appears to be a simple piece of glass used by two remote collaborators. Each person can draw with markers or put documents on this glass—but they can also see through the glass to the other person. This allows two people at work on a common drawing, diagram, or document to communicate with each other from across the world. Instead of looking solely at each other's faces or at a document (as you would through a screen share, or while working collaboratively with Google Docs), they look *through* their screen and into each other's spaces, as if through a common piece of glass. Cleverly, the image they're working on is flipped, so both can read it. Remember the scene from *A Beautiful Mind* where Russell Crowe is writing equations on a windowpane? Imagine someone else on the other side of that pane, co-creating with you (math genius sold separately).

One of the most interesting insights from Ishii's research generally was that a lot of communication is delivered through the gaze vector—the imaginary line extending from someone's eyes to their object of focus. Teleconferencing sucks because you can't watch people's gaze vector; you can't see what someone else is seeing or observe the information they're looking at. This visual disconnect is one of the contributors to the now-pervasive feeling of Zoom fatigue.

If you observe me looking at something and it holds my gaze, then you realize I must be interested in it, even if I don't say anything or point at it. This is especially true if a micro-smile flashes across my face or my eyebrows briefly raise. Conversely, if I see you frown and look away, the target of your expression tells me a lot. This process is subconscious for both participants and takes only microseconds. But seeing these glances, dwell moments, and the associated

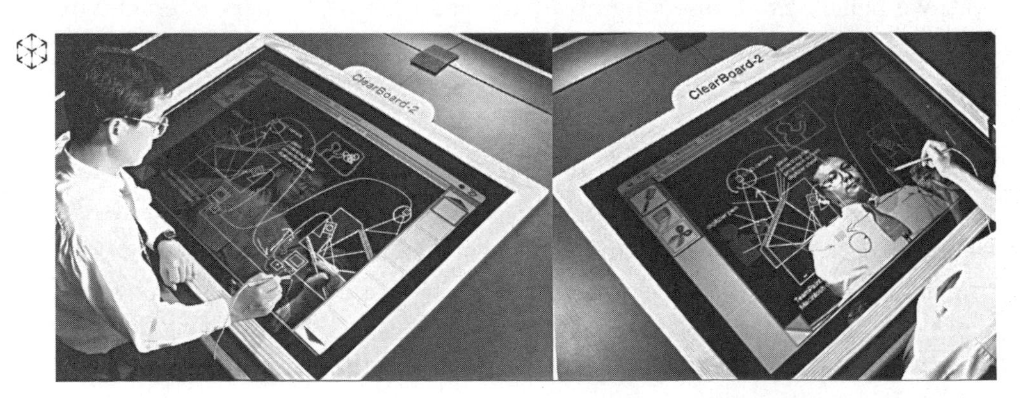

ClearBoard 2, developed by Hiroshi Ishii at NTT Labs, allows face-to-face collaboration on the same canvas with a view of the other person through the diagram or sketch.

microexpressions is what allows us to collaborate so efficiently when we are together in the same room.

Consider other collaborative decisions that you nearly always make in person that feel impossible through a screen. For example, have you ever been looking for a house to rent or buy, and FaceTimed your partner to do a walkthrough? You spend more time showing them your feet than you do the cute tiling in the bathroom. When you wander through a potential new home side by side, however, you get the chance to see your partner glancing at something, and can immediately glean how to steer the conversation to address their concerns and respond with more intelligence.

Today's video-conferencing systems don't offer this. We typically have two choices: seeing faces looking at and reacting to us, or seeing whatever our conversation partners are seeing. Most of the time, however, it would be more useful to see both of these things simultaneously, without having to switch between views. This is what the next generation of spatial telepresence should be aiming to do: "Hey Leah, you'll never believe what I'm looking at right now . . . oh wait, you can see it!"

Jinha Lee, a former student of mine and a protégé of Hiroshi's at MIT, recently started a company called Spatial to focus on this telepresence opportunity. Their product allows multiple people to teleport to a common space. Each participant wears AR glasses or uses a tablet, phone, or desktop computer, through which they view the shared space—a team room, perhaps, where Post-its, sketches, renderings, or design artifacts are arrayed on surrounding walls, and there is a 3D object like a product design or a car model in the center. If one person changes the object in front of them, it changes for everyone. Most importantly, because everyone also sees 3D avatars of everyone else, participants can observe what every other person is looking at, working on, or gesturing to as they speak.

Spatial makes flat PowerPoint presentations a thing of the past. It supports multi-person authoring and editing, spatially posting ideas, and clustering like any real-world Post-it Note workshop interaction. There is also a "vocal search" where you can whisper into your hands, and images of whatever you say fly out and land on the wall. And since you're in 3D, you can drag in 3D models or animations at any scale.

Spatial.ai creates a shared immersive experience using 3D-generated avatars for each participant.

You can also choose to show participants as 3D avatars, like in a video game. The system quickly molds each attendee's visage from a face scan so that everyone has a 3D presence, location, direction they are facing, and set of observable gestures. Doing this means that everyone's activity is *visible* to everyone else. This is a very different condition from the multi-faced Zoom call.

This 3D embodiment is the critical missing piece of telepresence and might finally reduce our need to fly around the world to effectively collaborate. If we can see others, see their work, see their microexpressions, and observe the subject of those microexpressions without latency, it will radically transform work. It will be as good as being in a conference room on the other side of the world—even better, when you consider that you get to do it without health risk, jet lag, bad airplane food, and dumping CO_2 into the atmosphere.

Working in a shared digital space also exploits the way spatial computing lets us appropriately scale our work both physically and temporally. Many types of work would be made easier if you could make things bigger or smaller, or faster or slower. A design review for a new engine is easier if you're Ant-Man, and urban planners thinking about traffic flows would love to be eagles aloft a thousand feet above the town.

SuperSight lets us create the ultimate collaborative design studio. For instance, I've always wanted a project room with a memory. One where I could scrub through time, or see a time-lapse supercut view of progress. Imagine if all the Post-it Notes, diagrams, sketches, and renderings covering the walls were attached to the debates and conversations that sparked them. Tap a sketch to see a projected hologram of the most salient bits of the discussion that prompted the sketch, the mini-pitch presentation of the ideas on which it hoped to elaborate, and layers of client feedback. If a teammate or client has been away from the project for a few days, the SuperSight-enabled team room might spotlight what had changed since they were last there, and which artifacts had garnered the most engagement.

This kind of virtual coworking technology opens up a whole range of collaborative possibilities. For example, the most natural and intimate interactions happen outside of formal meetings. (This may be why politicians play so much golf.) I recently went to an event at a ski mountain in Vermont where investors sponsored lift tickets for entrepreneurs to mix and mingle and do single or multiple four-minute pitch sessions as they shared rides on the lifts back up the mountain. The time-boxed moment was a quick way to get to know a real person in a casual context—professional speed dating with chairlifts. It was entertainment, social, exercise, work, and nature. That's what I want my future of work to look like: more skiing meetings.

How could SuperSight help us get there? We could choose to bring people into our field of view as we dip in and out of tasks at work, and as we need others' expertise or assistance throughout the day. Instead of scheduling and waiting for a meeting time, we could pop in and out of short telepresence sessions while fixing a snack, waiting in line, or taking a walk. Pitch sessions and standup meetings would become more fluid, casual, and spontaneous.

What would work look like if you could teleport that easily? Given the option, many people will avoid the daily subway or commute and work most days from somewhere more pleasant, as millions have learned to do during COVID-19, or as millions more with disabilities are forced to do because conference organizers often fail to provide accessible stages and podiums, event spaces, or networking venues.

Still, for intense design collaboration, or for meetings like conferences where sociability is the primary goal, in-person interaction is irreplaceable. Yet there's a downside to this ease of connection. Once we seamlessly weave ad-hoc staccato meetings into a now nonscheduled day, with an expanding constellation of people and projects, how do we differentiate between focused time and available time? Between personal life and work?

6.4 Forever available

continuous partial attention *work/life blur*

The experiences described so far are becoming woven into and are psychologically indistinguishable from work. We aren't *at work*, but we are *on work* persistently. For the professional class, there will be even less of a divide between the "placeness" of work and home, as we psychologically toggle ever more seamlessly between working and not working. With teleportation, the boundary between work and play will blur even more than it does now. You already take your work home with you in your pocket—soon you'll be able to see it through the SuperSight glasses you're always wearing, too.

I know it's a bit dystopian to say that soon we won't have any division between work and home. But every new telepresence technology pushes society further down this slippery slope. Let's face it: you're probably going to keep working once you've finished "working," anyway. So why not use your whole house, instead of your laptop? If you're a city planner, maybe the CAD designs you've been working with become the floor of your living room. If you're a commercial illustrator, maybe your hallways contain projections of your clients' projects. Instead of checking Slack on your phone while at the playground with your child, important notifications might pop up around the monkey bars or appear in the sky over the soccer goal. (Would this be *less* distracting than having your head down in your phone, or more?) SuperSight will give us this option to blend and mix work conversations and tasks into our current environment regardless of where we are or time of day.

The converse will also be true: instead of bringing your work life home, your home life will be feathered into work. Instead of checking a video feed of your kid in preschool on your laptop between meetings, they will be projected onto a wall of your team room, so you and your colleagues can see them playing right next to you, while your teleported dog rests comfortably at your feet.

With your dog and kid at the office, and your work at home, there's no primacy of place to regulate our focus and attention to "projects." We will check in with either work or home with the frequency of bathroom breaks. Toggling and fluidity become the norm. Coworkers will better understand your family and stressors, your partner will have more context to help solve work issues, and your kids will be able to vividly describe what you do all day.

When projected reality makes home work and work home, how will people know when others are "on work" versus "off work" and interact accordingly? When you can see a person in front of you, eyebrows furrowed, deep in flow mode, it's pretty obvious that now is not the time to tap them on the shoulder. How to determine when another person is and isn't accessible is one of the biggest teamwork challenges we face. In a future where we're always cycling in and out of work states, and interacting more frequently with partners and kids, knowing the appropriate moment to "drop in" becomes critical. How might we create serendipitous interaction and unscheduled collaboration when working at a distance?

We all have different productivity curves, which vary with the time of day, our mood, the weather, the number of hours of restorative sleep we've had, and so forth. How do you infer someone's interruptibility? While many current communication apps have status toggles, they aren't often used and therefore tend not to be respected. With SuperSight, however, status could be sensed automatically. Your glasses will use eye tracking to see if you're focused and in flow, or actually in need of stimulating conversation and would benefit from a phone call with a friend. Of course, this status information could be next to your name in Slack, but it could also be rendered as an interruptibility aura to signal your openness to conversation, the way the tops of NYC taxis signal their availability for hire.

Once we have this framework set up, it'll also further support those virtual drop-in interactions that spatial computing enables. If you want distraction or have some spare time, a system could connect you with someone else looking for a chat around the watercooler, or remind you of a mentor you should have reached out to, or alert you to a project in need of help that you might like to weigh in on. You're just going to use that blank time scrolling through Instagram otherwise; you may as well make it more productive. Take the time when you're already distracted and do something better with it. Such a system would also allow us to get in touch with supremely busy folks more easily. Everyone has downtime—some people just don't know when it's going to be.

Of course, there is a downside to losing downtime. American culture especially seems to fetishize a multi-tasking, thin-sliced, hyper-optimized lifestyle designed to squeeze productivity out of every idle moment. Linda Stone, a Microsoft researcher, has called this the problem of *continuous partial attention*. When we never fully focus, our performance suffers at work, and our relationships suffer, too. Smartphones have already sabotaged attention. Today, glancing at one's cell phone is authorized, normalized, and a ubiquitous gesture in our lives; we immediately reach for them in the slightest moment of distraction. Most of us can't even hit a button in an elevator and go up without fondling a phone. But what if, instead of getting through a level of Candy Crush, your phone instead prioritized strengthening a social connection, like sending a "Marco Polo" to your mom? You could even have a loose scheduling system based on the space you're in: you might code the system to connect you with a random overseas friend every time you hit the toaster button, or always connect you to your daughter as you pop into your self-driving car for a relaxing, reclining commute.

SuperSight has the potential to reshape the way we work. Given the right design choices, it can reprioritize relationships, filter distractions, and communicate to others that we are "in flow" and not available for the next seventeen minutes, but would love to teleport into their problem zone, or share a summit moment, after. And because SuperSight wearables can sense so much more, as we discussed in chapter two, they will be more attuned to our mental state, encouraging socialization and time with the family when we need them and they need us, or just moments of respite and humor that keep us human and healthy. They'll help us better navigate our work lives physically, socially, and psychologically, too.

Part 3

Society Scale

Now, let's zoom out to examine how spatial computing might impact larger societal issues: public health and safety, firefighting and other first responders, how city planners communicate and ask for community feedback, and more. We'll confront critical issues of access, and explore how the world might change for the better if we could see clearly into the future.

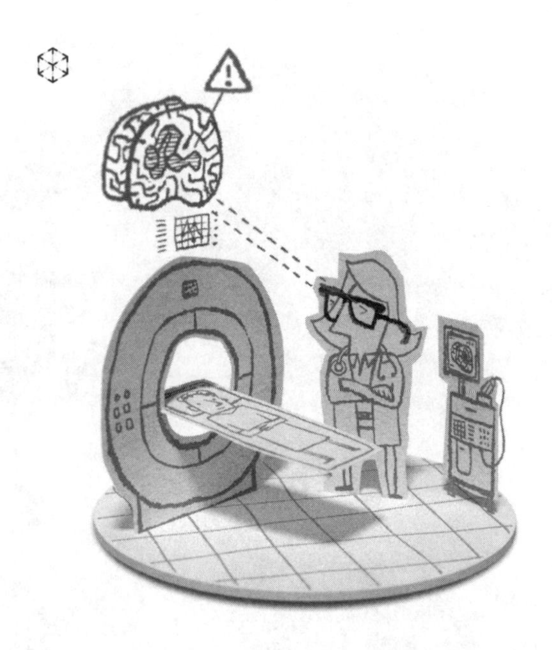

What if an algorithm could see better than your doctor?

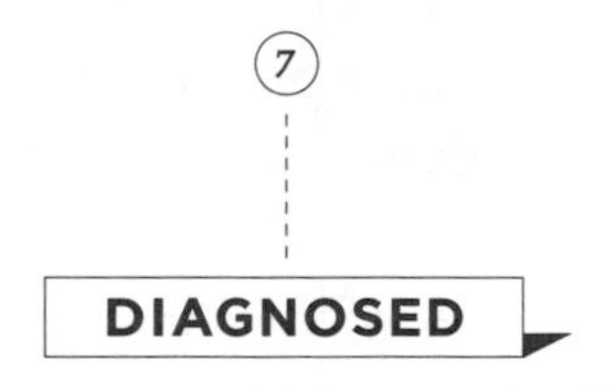

About healthcare, medicine, and AI nurses

7.0 Digitizing Dr. Rose

computer vision and public health

One of my dad's many talents was the ability to diagnose at a distance. Dr. Rose had the ability to spot people with kidney disease from their skin tone and kids with genetic disorders from facial proportions and symmetry. Once when we were sitting in a restaurant, a guy with an odd gait walked by. My dad nodded at him and whispered, "See that guy over there? He's got herpes."

I was impressed. "How can you tell? He's thirty feet away!"

"Look at the way his foot drops. It's a signature side effect of that disease."

Today, technology is enabling a world where the same kind of pattern recognition my dad performed can be trained into a deep-learning neural network to diagnose many conditions through visual analysis—not just kidney disease or herpes, but skin cancer, glaucoma, neurological diseases, and more—from photos, thermal cameras, radiology scans, and short-wave radar. Sometimes these diagnoses might require specialized camera systems, but other times the cameras in your home, smart glasses, and city streets will be enough to identify and diagnose illness—without a doctor visit.

The need for contactless diagnosis became starkly apparent during the COVID-19 crisis. Suddenly every country, transportation hub, and gathering place needed to establish policies and procedures to keep clientele and staff safe, and increase customer confidence that these policies were being followed. Some countries—starting with China, where the disease first appeared—began installing temperature sensors in airports and grocery stores and outside clinics as a screening tool. And as the world began emerging from quarantines and lockdowns, with vaccines still rolling out, many gathering places sought ways to perform passive, contactless, automated health checks before people entered.

At Continuum, we designed a human-centered health-screening solution that can be installed in any doorway, limits the spread of disease, and increases public confidence by signaling visually that "everyone here has passed the test." While our system was designed with COVID in mind, what I'm suggesting here is a more continuous, ongoing infectious health screening—something more akin to airport metal detectors, highway speed cameras, or parking lot security cameras that flag suspicious behavior like someone checking multiple cars. It's a public health tool for schools, corporate buildings, transportation hubs, restaurants, hotels, concert and sports venues, conferences, and retail stores. To be effective, this new type of screening door needs to do two jobs: detection and projection. By detection, I mean diagnosing if you're ill or not. And by projection, I mean including visual signals to communicate "health safety" to people wary about entering—like a ZAGAT rating for health.

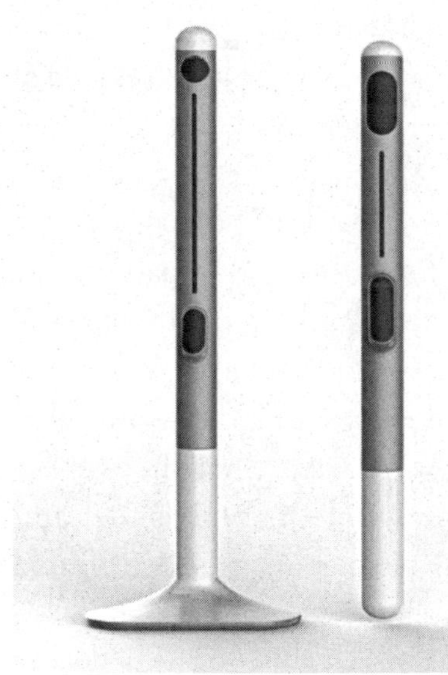

Responding to COVID-19, at Continuum we designed a thermal and X-ray screening tool for our clients: hospitals, restaurants, retail stores, entertainment venues, and transportation hubs.

In our prototype, we used a camera with computer vision onboard to isolate people and faces, then an industrial thermal sensor to take forehead temperature readings from meters away. If the system detected elevated body temperature, an X-ray sensor made a second reading to evaluate heart and respiration rate, both of which are also elevated in people fighting a virus.

Our wish is to keep the system anonymous and passive with a bias toward assisting rather than shaming or incriminating. When the pandemic struck Wuhan, Chinese authorities deployed drones with thermal cameras to spot fever from the sky, set up block-by-block barricades and checkpoints, and stopped people on the street to scan their foreheads with a thermal gun—a disease version of stop-and-frisk. Of course, this police-state approach wouldn't work in America. But SuperSight offers other, less draconian ways to keep people safe, to help those truly in need, and to generate confidence in gathering places. We don't want our new world to look like *Gattaca*, the 1997 sci-fi movie featuring turnstiles that take a small blood sample as you pass through to verify that you are in the genetically modified majority.

Continuum's design preserves anonymity, and isn't recorded. Rather than being punitive, it provides subtle praise for "passing the bar," like the sound of a hotel door unlocking or a Disney MagicBand flourish, rather than embarrassing red lights and shrill shoplifting alarms. Further, it provides assistance if needed, such as offering PPE, appointments with healthcare providers, or just a ride home. Our design philosophy is to use the least amount of personal disclosure necessary to ensure a gathering place is as safe as needed. In addition to sensing, the design serves as a visual beacon that signals safety and provides assurance for those people entering. While passive health sensing may sound overzealous or overreaching, this functionality would become as normal as handing over your ID to get into a bar.

New technology offers exciting ways to keep us safe and healthy with passive sensing at a distance—not just when it comes to dangerous viruses, but other health conditions, too, from skin cancer to liver disease and diabetes. You may not have thought of drive-through restaurants, concert venues, or churches as a place for healthcare, but these high-traffic places may be the perfect place for pop-up "micro-clinics" with passive screening and rapid testing, all powered by SuperSight.

7.1 Discerning from a distance

self-diagnosis | *home testing* | *incidental public testing*

Even before the COVID crisis, large technology companies like IBM Watson and Philips were rushing to apply computer vision to a range of remote healthcare applications. Startup entrepreneurs like Susan Conover got there first.

I first met Susan, then a PhD student at MIT, through an angel investor at an MIT mentoring meetup. She recently graduated from the Sloan MBA program—and had personal experience with undiagnosed melanoma. Spotting this condition early is critical for treatment, but many people avoid seeing a doctor for fear of embarrassment. This lets disease progress, and whether a rash is something harmless like eczema, or more serious like a sexually transmitted disease, it becomes worse than it needs to be.

So what if you could get a ballpark assessment without having to disrobe in a doctor's office? What if you could snap some photos in the privacy of your own bathroom to identify if your skin flare-up is just a reaction from a new soap or dryness, or something for which you should seek professional help?

To aid other people like her, Susan teamed up with another computer vision scientist to help people diagnose skin diseases. Her application, LuminDx, allows you to do just that. Photograph the suspicious rash, and the app's neural network processes your picture against millions more like it. (This takes about a sixtieth of a second. Imagine how much longer it would take your dermatologist!) The app displays the top matches of what you might be suffering from with other examples for visual comparison—or tells you it's likely nothing at all. If there's a match, it then recommends the right home testing kit, an over-the-counter medication, or tells you to visit a dermatologist, pronto.

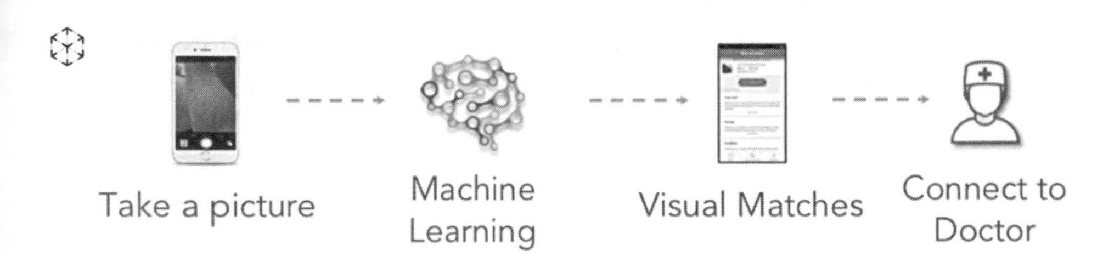

Identify skin conditions instantly with artificial intelligence using smartphone pics.

LuminDx used thousands of labeled examples of rashes and melanomas from the VA archives to train a neural network to help people determine if they need an at-home test or a dermatologist.

Susan has a typically Irish complexion: red hair and fair skin that is susceptible to sun damage. According to Susan, her doctor didn't spot early signs of melanoma or prepare her to see the skin cancer. She said he only cavalierly quipped, "'If you see anything uncommon, let us know.' But how was I supposed to know, as a twenty-two-year-old, what was common and not?" Fortunately, her mother spotted a new mole on her back one day at the beach and insisted Susan visit a dermatologist, who identified the cancer early enough that she could receive treatment before it metastasized.

While a graduate student at MIT, Susan dove into how computer vision—specifically a technology called "similarity search"—might help people and doctors identify visual patterns. She started interviewing primary-care providers and learned that they only spend one week in med school learning about dermatology. "PCPs see six to ten patients with some dermatology issue every day," she told me. "Half of these are misdiagnosed." For example, when a person visits the ER for Lyme disease, it is almost always diagnosed as psoriasis. This results in a steroid prescription, which gives a fungal infection fuel to spread more aggressively. Susan hopes that LuminDX will become, as she puts it, "what . . . the Hippocrates app has become for drug interactions: a reference tool that doctors consult every day to help them in their practice. A daily clinical-decision support tool."

Her business model has taken an interesting turn since I joined the company as a computer vision advisor. Originally the idea was to offer a consumer tool that displayed the top most-likely matches, then recommended suitable at-home testing kits. These kits typically cost consumers between $20 and $300, and whenever LuminDX recommends one, it gets a referral fee of $5 to $25 per patient. Now Susan has discovered another valuable revenue source: clinical trials. Pharmaceutical companies typically need 3,000 patients distributed across around fifty sites for a phase 3 trial. Each day of delay in recruiting patients for a potential blockbuster can be worth tens of thousands of dollars. LuminDX solves this recruitment problem by finding many more people with specific dermatology indications who might be eligible.

Susan's startup exploits the phenomenon of Dr. Google, where more than a third of people search their symptoms before booking an appointment. Except instead of Googling for "strange spots on my . . ." and relying on your own eyes to determine if you might have gonorrhea (assuming you can bear to look at the images), a nonbiased computer can assess that for you based on millions of verified images. If you're going to be a hypochondriac, you might as well be an accurate one.

Of course, there's a chance you could be misdiagnosed. But false positives aren't necessarily a bad thing; you'll then be prompted to go to an actual doctor, where you'll be put at ease that the AI was wrong. It's false negatives (when the system says you're fine and you're not) that we need to worry about. That said, when the medical issue is one that many people are too shy to see a doctor about in the first place, such as STIs, the chance a potential patient will take action outweighs the risks: an AI doctor is better than none.

Along with home test kits, using your mobile phone to do personal preliminary medical testing is decentralizing healthcare from the hospital into the home—or the workplace, car, or bathroom. From toilets that perform regular stool and urine analysis to pro-privacy cameras that detect balance, gait, and fall risk, SuperSight promises to make diagnostic tests more accessible and affordable, and less arduous.

Many companies see the potential for what I call *ambient healthcare*. Healthy .io in Israel recently began offering a computer vision system for people with diabetes to test urine remotely and send the data to their doctor and patient record via their smartphones. DeepMind in London (now a part of Google) trained a neural network to read retinal scans taken during an eye exam to diagnose early-onset diabetes, cardiovascular conditions, and, they claim, "more than 50 sight-threatening conditions." Streamlining in-clinic diagnostic tests is a great start, but to my mind, the transformational potential here is to roll out SuperSight-enabled ambient diagnosis within more casual, non-medicalized, incidental experiences, with city-scale impact.

When Warby Parker first recruited me in 2017, it was to build a new business of online eye testing using computer vision. The company was already known for their fashionable, affordable glasses, and for mailing customers five pairs of frames in a home try-on kit. But without an up-to-date vision prescription—something required by law every two years—customers couldn't finish an order. My job was to create an accurate home eye test that anyone could do, even from a cramped NYC apartment, with FDA-level rigor.

"Diagnosis at a distance is increasingly possible with sensors in our cities."

—John Brownstein, PhD, chief innovation officer, Boston Children's Hospital

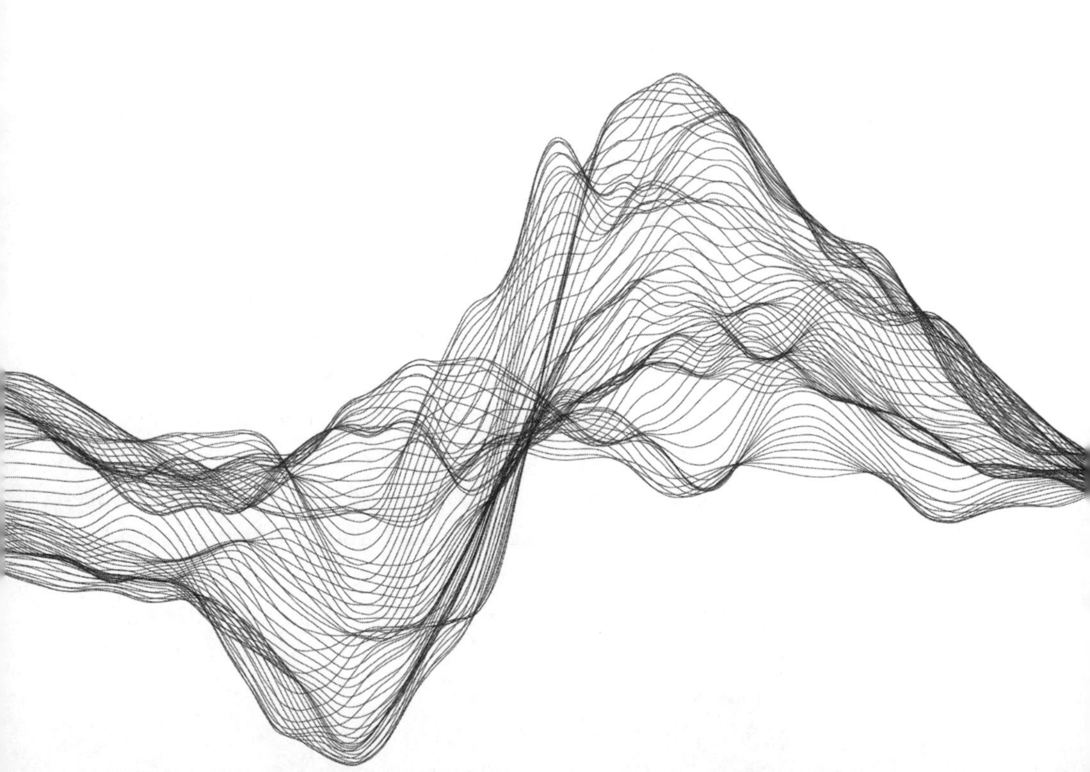

Warby Parker's home eye exam service also tests for astigmatism.

Vision testing—or measuring the eye's visual acuity, at least—requires only two things: optotypes, carefully designed letters or pictures of known sizes; and the ability to position a person at a precise distance away from those optotypes. Since many people have both a laptop or desktop and a smartphone, we used computer vision to do the distance measuring for us (who has a ruler handy anymore?). We created an eye chart of Landolt C optotypes (so called because they look like Cs rotated in one of eight different directions). The rows of the eye chart appear on your computer screen one at a time, from largest to smallest, and you swipe on your phone in the direction of each letter's opening. Then, within twenty-four hours, a real-live ophthalmologist licensed in your state looks at your data and health history and writes your new glasses prescription. It doesn't matter if your eyes are in Nairobi or New Orleans—as long as you have access to a smartphone, SuperSight can help screen your vision.

After setting up a new business of online eye testing in the US, and prototyping a mobile-only version of the same service, I wondered if there was a way to do vision screening in public places to democratize eye care, especially for kids. My team at Warby Parker started sketching experiences for high-traffic places like schools, libraries, science museums, playgrounds, parks, and viewing platforms on skyscrapers.

A Landolt C optotype eye chart. Landolt Cs provide a more precise test than tumbling Es because there are eight possible positions of rotation, and they work for kids who don't yet know the names of letters.

In 2017, I read a research paper in *Nature* that provided a eureka moment. Researchers at the College of Optical Sciences at the University of Arizona had built a compact, low-cost, auto-phoropter (the many-lensed "Better, or worse?" device through which you read the eye chart) to perform eye tests in developing countries. They combined two technologies to determine a corrective prescription: one to evaluate distortions in how structured light bounces off the retina, the other a set of focus-tunable lenses that immediately corrected aberrations. First the device determined a person's need for glasses, then it showed them how much better the view could look with a personal prescription—all within fifteen seconds. If we could make focus-tuning the lenses cheaper, I realized, such a system could work in less than five seconds—at least in theory.

I proposed building an instant eye test embedded in sightseeing binoculars installed in high-traffic places—such as the top of the Empire State Building, on cruise ships, and at summits like Mount Washington—to serve many people per hour. As you peer through the binoculars, if you need a larger diopter or an astigmatism correction, the tunable lenses adjust to show you the view with this correction. Just tap your phone to save the prescription—plus get a link to Warby Parker's virtual try-on app to pick stylish frames.

A fast self-service eye test using wave-front aberrometry and focus-tunable lenses to sense and correct distance vision could be installed in high-traffic public venues like this one.

As you can tell, I'm optimistic about using this kind of ambient healthcare to make public health screening more accessible and widespread. While performing an objective vision test, the sightseeing binoculars could also do a sub-second diabetes screening using computer vision–assisted retinopathy. Just as my dad was able to spot jaundice or neurological symptoms from afar, there must be scores of screening tests that are possible by combining powerful sensors and SuperSight AI algorithms. We just need to decide when early detection outweighs the costs of false positives, then figure out how to deliver these potential diagnoses to people directly, without accusation or shame. Once we solve those design challenges, this SuperSight application has the potential to make all of us a lot healthier.

7.2 The AI will see you now

collaborating with algorithms | *sharing mental models* | *agentive AI*

While AI is democratizing visual analysis tools for the average person to self-diagnose a rash or spider bite, doctors, too, benefit from strong-sighted AI assistants that help interpret imaging. SuperSight is already solving problems in some of the trickiest and most high-risk fields—like oncology. In her *Quartz* op-ed "I Was Worried About Artificial Intelligence—Until It Saved My Life," Krista Jones describes the challenge medical professionals face when looking for signs of breast cancer like hers: "Imagine for a moment that you are a pathologist and your job is to scroll through 1,000 photos every 30 minutes, looking for one tiny outlier on a single photo. You're racing the clock to find a microscopic needle in a massive data haystack. Now, imagine that a woman's life depends on it. Mine."

Every day, pathologists face an overwhelming task: accurately reading a torrent of mammograms to diagnose the 250,000 women a year in the US who develop breast cancer in an effort to reduce the number of those it kills (40,000 annually, on average). Limited by time and resources, these medical workers often, tragically, get it wrong; a recent study found that pathologists accurately detect tumors only 73.2% of the time.

This failure rate suggests a role for computer vision systems, which increasingly perform better at pattern recognition than humans. Startups worldwide are training systems to read diagnostic scans, and the results so far are impressive. A 2019 study in the journal *Radiology* noted that "machine learning coupled with PET scans allowed doctors to spot Alzheimer's disease an average of six years earlier than traditional diagnostic methods."

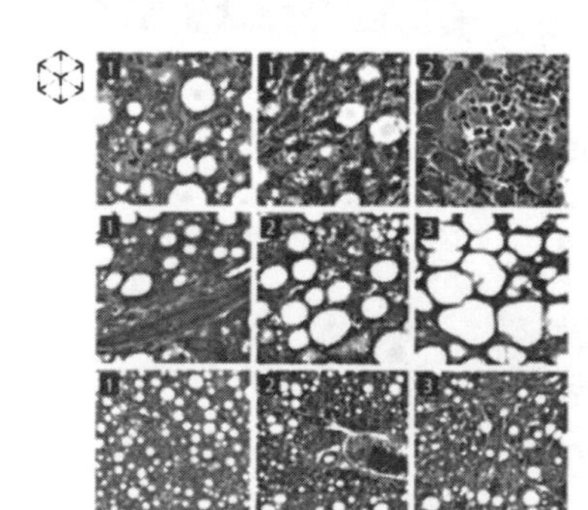

Example scans of liver cells analyzed with computer vision. The automated deep learning scores show good agreement with the human pathologist.

But if these algorithms are consistently as accurate or more accurate than human doctors, as seems to be the case, and far less expensive to employ, will computer vision put visual-pattern experts like radiologists and dermatologists out of work?

In 2017, prominent venture capitalist Vinod Khosla drew attention by predicting as much. "A world mostly without doctors (at least average ones) is not only not reasonable, but also more likely than not," he said in a much-discussed *TechCrunch* op-ed. "Eventually, we won't need the average doctor and will have much better and cheaper care for 90–99% of our medical needs." Many others agree with him, such as Andrew McAfee, co-author of *The Second Machine Age,* who believes that "if [SuperSight] is not already the world's best diagnostician, it will be soon."

I think this is naive. First, radiologists' and other doctors' roles, while augmentable by AI, will never be replaceable, because of psychological reasons like patient confidence or liability risks. Very few airline passengers would ever board a plane with no human pilot, even though autopilot technology already handles upwards of 90% of every flight. There are also functions that are so tied to uniquely human capabilities that they will never be automated: empathetic human interaction, for example.

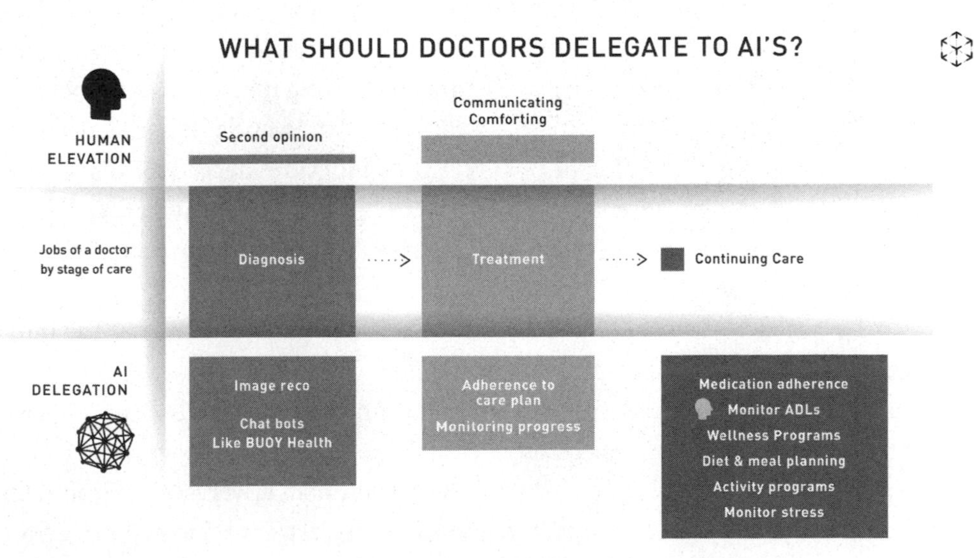

Algorithms are suited for predictable routines but high-touch interactions, empathy, and creative problem-solving are still the domain of humans.

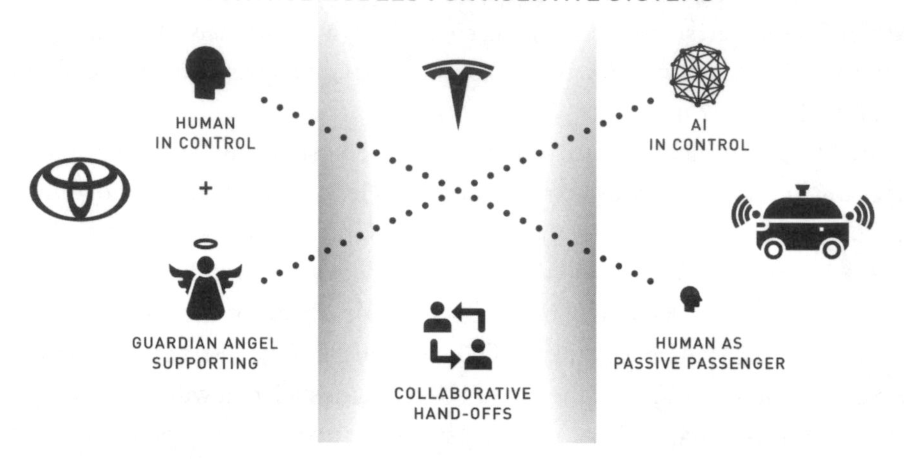

Especially for serious tasks like driving and medical conditions, we are always gauging how much to trust AIs, and our relationship to these new assistants. Should they only reach in and help before we make an error (passive), should they take over entirely for us (active), or should we hand control back and forth (collaborative)?

Second, it doesn't capture the nuances of how *jobs adopt tools.* New tools constantly change how we work, largely for the better. Computer vision will mature into a time-saving screening tool for medical professionals by streamlining workflows, improving diagnostic precision, and freeing medical staff to focus on other facets of their job, like research and patient interaction. It could also ease the workload of multiple doctors parsing over the same decisions. It's wise to seek a second opinion from another physician before a major surgery or course of treatment, for example. AIs and humans should both generate opinions, then compare and learn from each other.

AI assistance exists on a spectrum from active to passive. Consider the interface design in a self-driving car: Google has the tech hubris to put you into a car that doesn't even have a steering wheel. The control model here is clear. There is no human driving role in this equation. On the other end of the spectrum, Toyota's perceptual system is cast as a guardian angel. The human driver is leading, and the AI only "grabs the wheel" or "hits the brakes" at the last second to keep you from hitting a pedestrian or running off the road. But the most preferable control mode, for the near term, is at neither of these extremes: our Goldilocks, just-right Tesla.

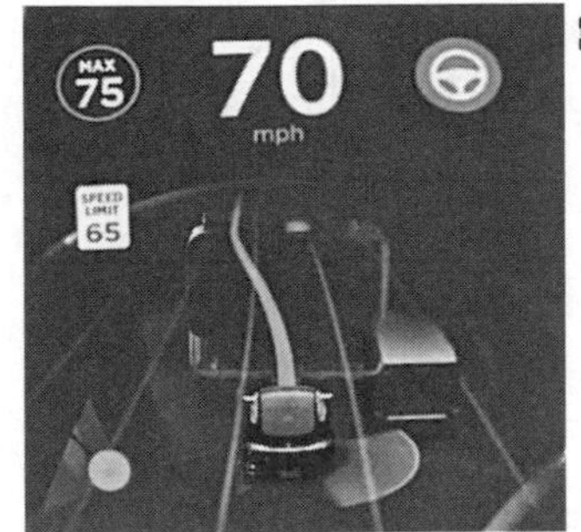

The Tesla dashboard dedicates a large amount of space to showing what the autonomous driving system can see and what it's planning to do next. Collaboration requires sharing mental models.

Tesla's self-driving car uses much of the space on its dashboard screen not for a big speedometer dial or tachometer, but to communicate to you what it sees (the cars ahead of you, the truck to the side, bikes, speed limit, or stop signs) and its plans (e.g., an automatic lane change to pass a slow-moving car ahead). The human's role here is akin to a manager's or supervisor's: they make sure the job is being done right. The driver, like an enlightened manager, knows the conditions in which their "employee" is likely to fail and may position a hand close to the wheel while driving over an icy bridge, through a construction zone, or while passing a large truck buffeted by the wind. Most of the time, though, this cautionary posture doesn't require people to take back control.

So what's required in order to structure an AI + human relationship where control is shared? For driving or reading X-rays, the needs are the same. The AI assistant needs to communicate its abilities and its level of confidence in executing those abilities, so that a human can determine when they need to step in or support, and when they can just let the AI go and do its best. The way this confidence is communicated in a Tesla is a large part of the dashboard display. How confident is it of the lane lines that it sees? Very confident is solid blue, less confident is gray, and a ghosted gray line on one or both sides of the car means that it may require your help, if there isn't another car to follow in front of you.

Think about an equivalent for radiology. How would an AI communicate its level of confidence in finding a hemorrhagic stroke? How would the system quickly communicate the need for a second, human opinion?

Our future AI assistants *will* require our attention. Sometimes that'll be useful, and sometimes it'll be distracting; sometimes they'll make mistakes that lead us to take our eye off the ball. In the case of radiology, if a system finds something that looks interesting to it, it may draw your attention away from the suspicious area you were watching—leaving the true source of the issue, which the AI didn't see, sidelined. In this way, there's a risk that an overreliance on an AI will decrease the performance of a human, in the same way that overreliance on spell checkers can lead you to use the wrong (if correctly spelled) word. The way forward will be to keep an eye on these assistive AIs and not lose our sense of intuition.

A period of adaptation in the medical field is unavoidable, as doctors learn to prioritize tasks that humans do best, such as creative problem-solving and establishing relationships with patients, while AI does the rest. The relationship between doctors and AI assistants will have a learning curve. But in the end, we'll all be better—and healthier—because of it.

⚠ HAZARD: TRAINING BIAS

As humans, we often trust too fast, and forgive too slowly. Believing the diagnoses that computer vision suggests means trusting our lives to autonomous systems. Yet we're rarely given critical information on how these systems were trained or how accurate they are. And what we don't know, in these cases, can absolutely hurt us.

Medical mistakes are one of the leading causes of death in the US. And it's inevitable that, as more AI-trained systems are used in hospitals, we'll start seeing more errors—potentially fatal ones—attributable to algorithms, even if the total number of errors is dramatically improved. Who do we blame for these errors? When there's an error during a surgery, the surgeon is held responsible; when a piece of hardware malfunctions, the device manufacturer is the one who gets sued. But when it's a line of code that doesn't do its job, do we assign responsibility to the engineers who coded the algorithm, the doctor who supervised the diagnosis—or, somehow, the AI itself?

Medical bias is another issue that doesn't go away just because you add AI. Algorithms only repeat the errors or truth found in their training data. And in the case of medical research, that unfortunately means datasets of white, middle-aged Western men, and young, plucky, privileged medical students looking to earn an extra buck by allowing themselves to be prodded for university studies. Neither of these groups is representative of the population at large. This is why we know less about how certain drugs interact with people of color or women, and why we are less versed in how the symptoms of the same diseases may present themselves differently in people from different ethnic backgrounds.

For example, we only realized in the past few years that women show quite different heart attack symptoms than men. The reason? Nearly all of the data on heart attacks was compiled from middle-aged white men. Even when it comes to breast cancer—arguably one of the most studied cancers out there—nearly all of the data is pooled from white women of European descent. This despite the fact that what little we *do* know suggests that African women have a higher chance of having the BRCA1 gene, develop more aggressive forms of breast cancer, and are twice as likely to receive a breast cancer diagnosis before the age of thirty-five.

If SuperSight is going to have a positive impact on how we diagnose disease and craft treatment plans, we need much more than algorithms. We need to fundamentally redesign the way that we're studying and treating disease. But just as important, developers of automated systems must take extra care to curate training data to minimize racial and ethnic bias, and create processes to keep humans in the loop to ensure data quality and transparency in algorithms. AI is only as good as the data we use to train it, and "clean," unbiased datasets are the grand challenge of the decade.

Because if the systemic inequalities in the health system aren't solved, and AI-influenced medicine starts making the lives of minority patients even worse, it's not the algorithms we should be blaming—it's ourselves.

7.3 Predicting and preventing prognosis

activities of daily living *pose detection through walls*

SuperSight won't just be used to diagnose and treat existing medical maladies. It will also be used to notice potential issues *before* they become a problem.

As more cameras are sprinkled around our homes for security and entertainment, these same eyes will be used to survey our wellbeing. First these systems will build up a representative baseline of our "normal" over time—how we typically move around, how much we turn in our sleep, the amount of bounce in our step, how we gesture and the extent of our reach—so they'll be able to tell when something changes. Is your knee bothering you? Are you increasingly sedentary? Do you look a little jaundiced or pale? Is your neck stiff? (Whose isn't?) Computer vision can help sense subtle, early signs of depression, weakness, even trends in cognitive decline, then recommend proactive steps to avoid an incident or costly emergency room visit.

Healthcare isn't the first field to use computer vision to detect human activity—the technology was invented for casinos! (What are the odds of that?) Those sprawling halls of slot machines often deal with sprawling customers, who they need to assist off the floor and back to their rooms to sober up, or maybe just back to the nearest slot machines. In the 1980s, casinos took their CCTV streams from the gambling floor and fed them into computers using a rudimentary pose classifier—an algorithm that recognizes a human head, arms, and legs. These systems could sound an alert if a guest was lying flat on the ground. While they couldn't determine if the cause was alcohol, a fight, or fainting from winning a million dollars, in any case a security person could be dispatched to that sector.

If you've never passed out in Vegas, then you probably first encountered pose estimation algorithms while playing video games in your living room. The first wave of "exergames" (a portmanteau of exercise and games) were enabled by Nintendo's gesture-based Wii Remote and Balance Board in 2007. These accelerometer-based sensors that you hold in your hand or stand on made it possible to detect body movements in three dimensions. While this was useful for games, it also had serious health applications. Soon physical therapists started to use remote-motion exercise games to improve adherence to prescribed daily exercises, a well-known and difficult challenge. The motivational effects of these games were compelling; people recovered faster as a result.

Next, the Kinect was launched in 2010 alongside a set of video games that you controlled with your body. This camera system used the same method the Apple iPhone uses to unlock using your face: a grid of infrared dots projected onto the subject, which is then read by a dedicated infrared camera. This markerless motion tracking, which can recognize people and their movements, excited both physical therapists and researchers (like me). It's the holy grail technology for helping people age in place in their own homes without the financial burden of nursing homes and managed care, because it lets us spot early signals of depression or decline.

The contemporary versions of these cameras are increasingly able to sense activities of daily living—the industry term for the patterns that represent our day-to-day—and discern trends in these activities. They passively log whenever you get up more frequently in the middle of the night, steady yourself with a kitchen counter or door frame, or struggle to pick up something off the floor or reach a top cabinet. These seemingly small changes in your behavior—barely perceptible to your partner and kids, much less yourself—sometimes presage larger health events that can be addressed before serious symptoms appear or debilitating disease sets in.

SuperSight can do more than track your physical movements to recommend health interventions. As we saw with the health screening devices at the beginning of the chapter, sensors in the nonvisible microwave and infrared spectra can read your temperature, heart rate, and breathing from across the room, too. Determining these from a distance multiple times a day is useful in detecting psychological wellbeing; heart-rate variability, for example, can reveal stress.

But why would we need a camera to tell us when we're feeling sick or stressed? The problem is that we're predictably poor at self-knowledge. Without sensing technology and feedback, we usually don't recognize or register incremental change, especially as we begin to age. Growing older is a natural process, but it can be messy and embarrassing at times. Our bodies begin to fail us, whether gradually or suddenly, and it can be hard to tell when that's just the new normal or if intervention is warranted or justified. As more people (of all ages) are living alone today than ever before, you can see how SuperSight would provide comfort to individuals as well as their remote families and caregivers.

Using SuperSight to detect that someone has fallen is important and relatively straightforward. Predicting that someone *will* fall, before they are on the ground with a life-changing hip injury, is nontrivial, but also possible. Pose detectors and gait analysis can determine if someone is shaky, or not walking symmetrically, or subconsciously touching surfaces for stability, or has a diminished range of motion. Using these subtle movements, other algorithms can predict the likelihood and probable cause of an upcoming stumble so people can get preventive care *before* shattering bones and enduring the trauma and risk of hospitalization. Falls are the primary trigger for families deciding that it is no longer wise for aging loved ones to live alone, so the potential here for improving people's lives is enormous.

Of course, houses don't offer unrestricted views of their inhabitants' daily movements; they tend to be full of pesky solid objects that normal cameras can't see through. This can leave mainstream computer vision largely powerless in cluttered spaces, even if your interior-design taste skews more toward Danish minimalism. Fortunately, AI-powered computer vision platforms are increasingly able to see through furniture and walls.

Computer vision PhD and entrepreneur Alexandre Winter is one of the people pioneering the use of SuperSight's observational powers, first for counting cars and pedestrians in cities with an app called PlaceMeter, then for the home-security camera company Arlo (now a part of Netgear). Home-security cameras want computer vision not only to differentiate between friends, delivery people, dog walkers, and foes, as we discussed in chapter one, but also for more mundane recognition tasks—is that backyard movement a wind-blown tree branch, raccoon, or intruder?

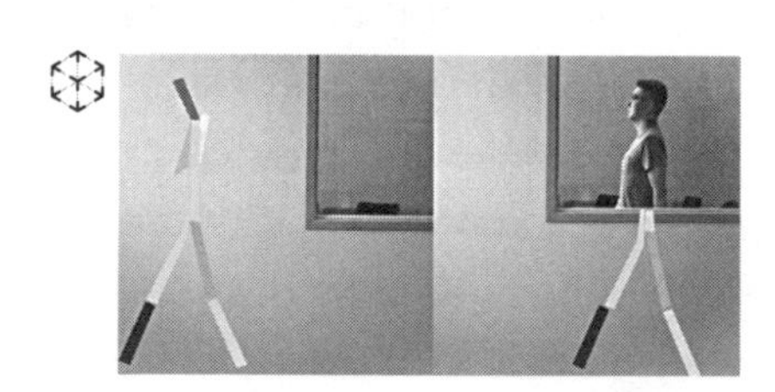

Pose detection and activity classification through walls using short-wave radar and machine learning.

Now Alex is using his SuperSight expertise to go after the $3.7 billion market for personal emergency response systems. Many seniors have a wearable device, typically a bracelet or pendant, with a button to press if they fall. Despite the fact that they pay a monthly service fee, usage is spotty. The devices are awkward, require recharging, and are seen as an emblem of frailty by their owners. Alex's service, Norbert, can see through walls to detect falls so wearables aren't necessary in the home. It's the SuperSight version of "I've fallen and I can't get up."

The Norbert device has four sensors, each trained to see a different range on the electromagnetic spectrum. A visible-light camera uses facial recognition to recognize who is in the home, an infrared sensor determines body temperature and heart rate, a millimeter-scale radar sees through clothing to measure the rise and fall of the chest, and—the most impressive feat—a longer wavelength, centimeter-scale radar detects people and their poses through walls.

How would you feel about being monitored by a system like this? It probably depends on how you feel about who is doing the watching, whether any information is recorded, and what you are doing while you're being monitored. How might we offer the benefits of this kind of SuperSight "guardian angel" service without privacy concerns? The solution is to use edge computing, as described in chapter one: the video analysis and processing would be done locally, at the camera. This approach means no human would ever see any video frames, except those uploaded if there's a suspected fall, and no cloud server would ever store any of your images. Yes, the camera might be watching you sleep, but it only uploads an alert if you seem to be sleeping much worse—never what you actually look like in bed.

In the coming years, this kind of pro-privacy approach to a home safety net will become as standard as doorbells. SuperSight algorithms like activity detection will provide an ongoing visual portrait of your lifestyle as you prepare dinner, play RISK, and fall asleep in front of *The Late Show*. We'll get the peace of mind that comes with knowing someone is watching over us—without the Big Brother surveillance problem.

7.4 Quantified self versus quantified health

data, security, and health insurance

So far we've seen how SuperSight can be used to diagnose someone who is already sick and predict when someone is about to need medical assistance. What about using it to prevent them from getting sick in the first place?

In March 2020, Boston was hosting that year's TEDMED conference just as the coronavirus pandemic was accelerating in the US. At Continuum, we were thinking about how we could use pro-social design to make sure people wash their hands. A conference center is the kind of place where handwashing is especially critical: thousands of people have flown in from around the world, and all interactions start with a warm but transmissive handshake. We made a couple of SuperSight-enabled handwashing prototypes, then installed them in a men's room usable by attendees—including the surgeon general, who was one of the speakers—to get feedback from infectious disease researchers and doctors.

Here were the three top ideas we advanced to the prototype stage:

#1 Watchful Cat

#1 WATCHFUL CAT

The inspiration for the first prototype came from the insight that we exhibit our best behavior when we're observed. But no one wants a Big Brother camera observing them, especially in a bathroom. Being observed by a cat, we thought, might be more tolerable. Who is a kitten going to tell? The ubiquitous raised-paw Japanese cats provided inspiration, even more so when we learned that they aren't waving—that hand movement is actually them washing their face.

For our device, the cat sits on a platform containing a proximity sensor, a ring of LEDs, and speaker. As soon as the cat's camera sees your hands moving, it meows a twenty-second version of the *Jeopardy!* theme—the length of time you're supposed to wash your hands—as the LEDs progressively light up. By the end, you're both smiling and clean.

#2 PROJECTED GERMS

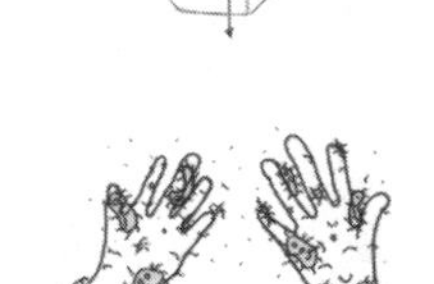

#2 Projected Germs

When do you wash your hands the longest? When you can see how grimy they are. I experienced this when I worked in a bike shop in high school: I would scrub my hands for five minutes after shifts to get the grease off. By making invisible germs visible and repulsive, we could show people the need to keep washing. Though some ultraviolet lights can reveal germs on surfaces, that wasn't going to work for the coronavirus—we needed to make the germs less abstract and more visible. So we hid a tiny pico-projector above the sink to shine really gross-looking microbes onto users' hands that transitioned to sparkles over twenty seconds. An extra advantage of this design over the previous one is that it keeps your attention focused on your hands, versus humming along with a cat.

#3 PETRI DISH COUNTDOWN

#3 Petri Dish Countdown

Iron Man has a glowing chest ring, the palladium source of his power. What if you could feel empowered by hygiene in the same way? In our third prototype, a pico-projector displays a glowing countdown on your chest, reversed so that it appears the right way around in the mirror. By using your body as a canvas, the device makes the information feel both personal and incriminating. We designed germy numbers that appear in a Petri dish to count down the germs you are washing away. (We tried to name and illustrate which germs are washed away in which order—*E. coli*, *Staphylococcus*, *Shigella*, the coronavirus—but couldn't get scientists to agree on a sequence.)

The only way to make great products is through experimentation with real people in real time, and I love the moment where we take "hardware sketches" (barely working, low-fidelity prototypes) and place them in the context of use. There is so much to learn in those first hours of interviews after your audience has been interacting with a prototype. This was also my first time conducting this research in a bathroom setting, surrounded by a group of guys with very clean hands. The primary takeaway was that most people do a terrible job of washing, but to do better, as one physician emphatically stated, "all you need is a twenty-second countdown timer in the mirror!"

Is it overreaching and intrusive to monitor handwashing? Not if it removes a primary vector for spreading a virus that kills. Nobel-winning behavioral economist Richard Thaler studies systems, like ours, that encourage people to behave in their own best interests. "Libertarian paternalism," Thaler writes, "is a relatively weak, soft, and nonintrusive type of paternalism because choices are not blocked, fenced off, or significantly burdened." We don't force people to wash their hands thoroughly for twenty seconds; rather, the Watchful Cat's twenty seconds of purring the *Jeopardy!* theme and twenty-second fill ring set an expectation people can choose to follow. It's not a demand; it's a nudge.

Systems that sense and reflect back to us our daily actions in a personal behavioral mirror, then nudge us in the right direction, have enormous potential. The design challenge here is to make the sensing passive (no effort or logging required) and the display or results unavoidable (no apps, please). SuperSight provides this passive sensing platform for behaviors, and augmented vision the ideal method for clear nudges.

Healthcare service designers and product leaders need to find the Goldilocks zone with SuperSight sensing and feedback for hundreds of daily health decisions—what and when to eat, bedtime routines, stress management, socialization and conversational scaffolding—nudging healthy activity and avoiding risk in the process. If the feedback is too creepy or too nudgy, people will reject the service and toss the device out the window. But if these systems are subtle

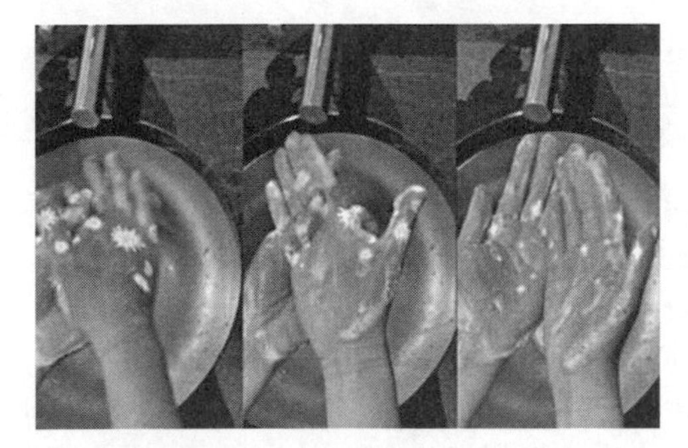

Projection of magnified, animated germy grossness to motivate better handwashing.

and persuasive, they can positively change behavior and bend the cost curve for all of the chronic disease categories, each of which accounts for billions of dollars in lost productivity and healthcare spending every year.

As so much intimate, real-time knowledge about our health and activities becomes accessible through powerful sensors like SuperSight, we must also use opt-in and similar policies, along with pro-privacy network architectures like edge computing, to safeguard personal information. This data, when used correctly, can encourage positive health behavior, but it can also lead insurance companies to refuse policies or charge extortionate prices for those with preexisting conditions or habits that make expensive care more likely. This concern around data sharing is one of the reasons people decline genetic testing or refuse mental health care. To mitigate this risk, we need to be transparent and specific about who has rights to monitor an individual's data (their family? Their primary care physician? Researchers?), for how long (over days or years?), and at what resolution (personal, or aggregated and de-identified?).

There are many good reasons to share your health data, and lots of technologies to do so anonymously. When it comes to public health, larger datasets across more distributed and varied populations create more robust and accurate science, and better predictions and courses of treatment for everyone, providing a way out of the data bias problem that currently plagues medicine. Apple is taking a leadership position with Apple Health and the myriad of data coming from the more than 50 million people who wear an Apple Watch every day. They collect and store information about heart rate, EKG, stress, physical activity, meditation, sleep, and handwashing, and then de-identify it for science and medical research projects.

As of this writing, America's health insurers can't legally charge more due to preexisting conditions—but they *could* use data to promote specific programs for, say, obesity, hypertension, or smoking cessation, and offer real cash incentives or discounts on premiums in exchange for demonstrating healthy habits. At first, this seems fair. Some health insurers already offer cheaper rates or Amazon gift vouchers if you can prove that you're doing the right things to reduce your stress and keep your body healthy. Have a gym membership? That might bring your monthly dues down a little. Regularly log your caloric intake? Bump them down again. This self-selecting, opt-in method of lowering your premiums is awesome—if you're on the right side of the wellness curve. Those who are not—whether because they have chronic illness, are temporarily disabled, or live in a food desert where it's harder to eat well—suffer as a result, despite being the patients who need health insurance most.

How could SuperSight factor into a world of dynamically priced risk? Well, the camera doesn't lie. Instead of you self-reporting your smoking or driving behavior, insurance companies' AI algorithms could observe it directly through

My son shows off a maker project to encourage handwashing: a timer that purrs a tune. We posted the project's design and source code on Instructables.com, where it was featured on the home page, to inspire others to make their own version.

virtual eyes in your car and home. Insurance companies are starting to subsidize the cost of quantified-self devices like activity trackers, internet-connected scales, glucometers, and blood pressure cuffs. Your smart watch, for instance, tracks how many steps you're taking, your stress level, if you meditate, and how well you're resting at night. Insurers will also process your public social media feeds to find your gym selfies—and, on the opposite side of the risk scale, your new obsession with cave diving.

While many people over age forty are wary of sharing *any* data with insurers, this tide is turning for the Venmo demographic, especially if insurance companies take an opt-in, two-step approach to data sharing. The first step is to let individuals see their data and trends; the second is to give them the choice, *only if they wish*, to share that data with others. We're now seeing this pattern in automotive insurance. At first, Progressive and other auto-insurance companies rolled out black boxes for cars that collected speed and acceleration data and uploaded it directly to their corporate servers. Customers totally rejected this. Now State Farm offers an app that collects the same data, but presents it as a safe-driving score to the individual, who then can elect to take the optional step of sharing it with State Farm for a discount.

Health insurers should follow this playbook. First, offer a set of wearables, connected devices, and SuperSight-enabled sensors that can assess lifestyle, risks, diet, and mood; predict falls; sense an early cough; and more. Then, present these private results to customers first, and empower them to choose if and with whom to share their data, offering a tangible benefit in exchange. The resulting democratized data architecture for health and wellness information could revolutionize proactive and preventive care for individuals and, in turn, create a healthier and safer society.

What if a hologram could
help you do your job?

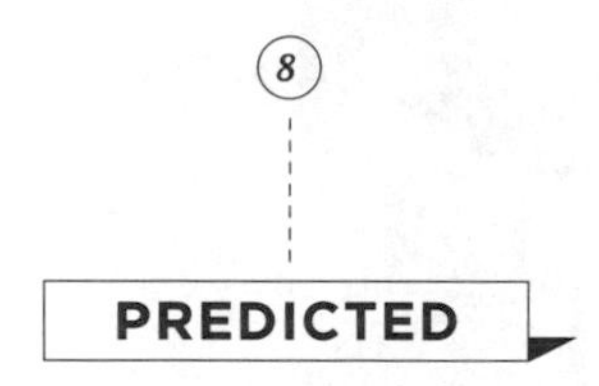

About industrial and city-scale applications of AR

8.0 Seeing the way through smoke

full spectrum vision for first responders

My brother-in-law Ed Poli is a real-life superhero. He served in the Peace Corps in Tonga, then worked at inner-city schools at Teach for America, and today is a fire captain in Harlem at one of the world's busiest firehouses. Every day, Ed and his crew run up tall buildings with sixty pounds of gear, break down doors, find the source of fires, and sometimes carry terrified kids down six-story ladders to save lives.

Last summer, we were fishing on Long Island Sound (Ed's other passion), talking about my work in augmented reality and X-ray vision for health care and construction. Suddenly his eyes lit up.

"Stop right there. You're saying we could make glasses that could allow us to see through this water?" He started to gesture excitedly. "I want to see shoals and deep spots! I want to actually look down through these glasses and see the school of porgies!"

We took a couple of casts, both imagining what the water would look like if it were as clear as glass: how we would see those porgies and lobsters below, and how the stripers would slice through at thirty miles an hour before—ideally—they hit one of our lures. After a few minutes of silence, he had an idea: "You know what would be a game-changer for my firehouse crew, David? Figure out how to let me see through smoke. That way I can get to people more quickly and to the source of the fire."

The Smoke Diving Helmet sees room features and people through smoke.

Ed's wish is now possible. A California company called Quake gives firefighters "bionic" eyes to "see through" darkness and smoke. Their Smoke Diving Helmet highlights outlines of walls and people, and shows thermal "hot spots" and vortex flows via colors. The helmet also masks the sound of the firefighter's own breathing with noise canceling, so they can better hear victims and others on their team.

Firefighting is information intensive, time critical, highly collaborative, and dangerous. Ed told me that as the captain of his fire company, he can request a basic description for any building in Manhattan on the way to the fire, but it'll be on paper and often out of date. He would love to give his crew more navigational help in buildings, especially after they have located people and need to determine the safest paths out. This data, known as BIM (building information modeling), is now all available in the cloud for buildings constructed in the last few decades. SuperSight glasses that could show such a map of the built environment would be incredibly useful to first responders, like the imagined gurney from chapter one that projects arrows on the ground to help EMTs navigate efficiently to patients. The ideal SuperSight firefighter helmets would show a projected map of the structure of the building, the position of other firefighters and their bio-stats, and the ambient heat flowing past (an indicator of how close they are to the fire's source). Drones with thermal cameras could add more "third eyes" for situational awareness as they circle outside or even fly through buildings.

The arguments for applying augmented reality here are irrefutable. A helmet with thermal cameras and telemetry embedded that makes emergency responders more omniscient, and better able to save lives, is low-hanging fruit. Few would argue against funding their design, development, and deployment. It's also a great lens through which to peer into the future of augmented reality.

In design research, we often try to find environments that represent extreme use cases. Designers go to industrial kitchens for insights on cooking at home, or interview mountaineers to learn about designing better coats. Applications in firefighting, construction, and industry are early indicators of how SuperSight's disruptive power may alter our working lives beyond the video game tropes we discussed in chapter six.

8.1 Superworkers with X-ray vision

see-through cities *digital twins* *AR clouds*

Whether they're fighting fires or fixing cars, millions of industrial workers will soon have augmented "bionic" eyesight. Instead of this feeling like a huge disruptive leap forward, though, it'll simply be an extension of standard work gear. Today, nearly everyone who works in a factory, power plant, refinery, or other industrial setting already wears safety glasses, just as a firefighter already wears a helmet. In the near future, the integration of SuperSight into these protective Plexiglas visors will allow them to learn new tasks on the job, check their work, and even get an expert to "see through their eyes" and provide help when they need it.

Imagine it's your job to repair an airplane engine, a transformer atop a telephone pole, or a pump at a nuclear plant. The task is complex, and for every hour of delay, millions of dollars may be lost—but if the job isn't done right, the consequences are catastrophic. Traditionally, you might consult a huge three-ring binder with hundreds of procedures for disassembling and reassembling components, running tests, consulting checklists, and so on. Soon all that information will appear in your field of vision instead. You'll look at the machinery you're fixing, and you'll see the type and model as well as that specific machine's maintenance history and performance record. Your glasses will guide you through each repair step, illuminating individual parts as well as actions you need to take. As you work, your glasses confirm each step is done right, recommend appropriate tools and techniques, and shine a spotlight on the next step. Your work might also be uploaded to the cloud and attached to a *digital twin* of the object to create a training model should another person need to execute the same steps.

A digital twin is a data-driven replica of an object (like an engine) or system of things (like a building) that includes information on how it was built and how it functions—a virtual representation of an object that bridges the gap between the physical and digital worlds. Think of it like creating a virtual double of your engine, airplane, or rocket ship. It contains detailed data about all the components, and a persistent log of interactions: who assembled it, last and next service dates, likelihood of failure, and more.

These digital replicas also become a safe, low-cost space for testing and iteration. A company wouldn't want to experiment on an operational multimillion-dollar production line, but by creating simulations, it can predict problems before they happen and test new ideas without physical risk or material waste. If the concept works in the virtual world, it can then be sequentially substituted into the real one.

The origin of twinning was at NASA, which made two of everything for the Apollo missions in the 1960s. When there was an issue with a rocket in space,

Digital twin data layered over an industrial motor reveals performance metrics and service instructions.

engineers used the twin on Earth to troubleshoot or figure out and test what parts could be repurposed, as in the now famous Apollo 13 near-disaster.

Back on Earth, the technology to generate digital twins is all around us. Google and Apple cars with spinning LIDAR (light detection and ranging) roof cameras started digitizing streets and buildings a decade ago; now all of us are working on the project to create a world-scale digital twin map of the built environment. Every photo you take and upload to the cloud includes a little line of metadata containing the latitude, longitude, altitude, and direction; photogrammetry algorithms are then able to use multiple flat photos of the same scene to infer 3D structure. The most recent phones have LIDAR sensors built in to capture 3D scans at sixty frames a second and centimeter resolution. Cars, with their multitude of sensors, are another type of scanning system that feeds the world-sized AR cloud. And when workers start walking through power plants, factories, and office buildings wearing glasses with range-finding cameras, we'll digitize the world even faster. Even your Roomba is in on the scanning game: as it's rolling around your floorboards, bumping into your coffee table, it's also creating a digital twin of your living room. That may be the most important job your iRobot vacuum cleaner performs: not re-spreading cat hair around your carpets, but mapping your furniture and decor with incredible accuracy—and sharing it with the AR cloud.

Yes, automatic digitization sounds a little creepy. But what we get in exchange is incredible—first for navigation around and through complex places like hospitals, then for inventory and insurance, then to predict risk, and perhaps to propose new designs. Many services will use this digital replica of the world to measure and model from.

We've seen how thermal cameras can extend our vision through smoke and fog, and how digital twins can superimpose 3D plans and performance data over machines. But wouldn't it be amazing to peer down through layers of concrete and pavement to see massive water and sewer pipes, subway tunnels—the essential infrastructure of a city? Thanks to a company called ESRI, an early leader in geographic information systems (GIS) software, the digital twin data necessary for such city-scale transparency already exists.

In the late 1990s, ESRI's desktop ArcGIS software was adopted by customers with geo-coded data like oceanographers, mining companies, transportation and city planners, statisticians doing political redistricting, scientific researchers, and

City water and electrical infrastructure is revealed as lines of light superimposed over urban streets in this visionary project with Microsoft HoloLens and GIS company ESRI.

more. Now all that geo-coded data is in the cloud and accessible through APIs used by literally thousands of apps on phones, tablets, and soon, smart glasses. Because the placement of water and electricity pipes can be hard to predict during construction projects, ESRI developed an AR app for Microsoft's HoloLens headset that allows utility workers to look down and see the city's circulatory system running under the cement. If they spot any discrepancies between the digital twin and the actual location of hydrants or streets, they can "true up" directly on site. The app's 1:1-scale overlay map of the city helps construction managers know precisely where *not* to dig when, say, installing a new bus stop, to avoid an electrical shock.

Two things are required to make this kind of projected map illusion work: a digital twin, and a spatial anchor—whether GPS coordinates or a recognizable stationary object like a building facade—that lets you position and scale the digital replica to match the actual object. Using this alignment mechanism, data can be projected directly onto an object whose guts you'd like to see. This gives workers the power to peer into everything from electrical equipment to heavyweight machinery, without risking the danger inherent in direct contact. The digital replica of an electrical transformer or milling machine is projected into your field of view, aligned and scaled perfectly to fit the object, and mapped to track the object as you move your head and walk around—it will feel like animated X-ray vision.

Imagine you're an engine repair technician. Not only will your glasses superimpose a 3D replica over the real machine, they also will spot broken or worn

"Our workplaces, factories, and power plants—the entire industrial world—should be decorated with augmented data."

—Jim Heppelmann, CEO, PTC

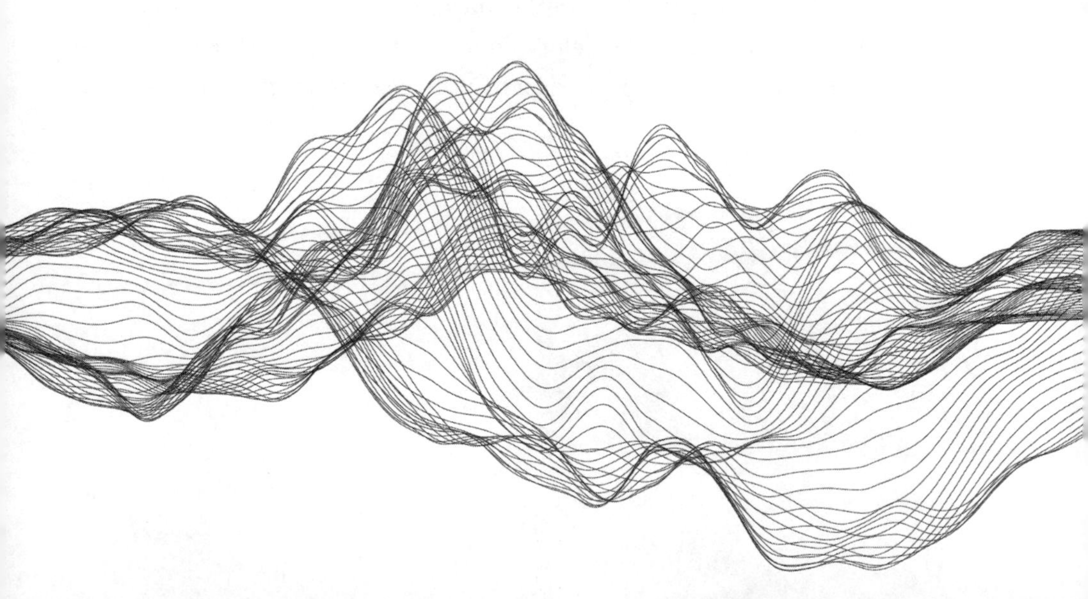

parts. They will magnify and perfectly illuminate your visual field, so that you can zoom in to look closely at wear and tear, or work on a very small part—like pinching an image on your phone to make it bigger, except in real life.

The system will know where past mechanics have made costly mistakes, and tell you where to slow down, rehearse complicated and risky steps, or get some expert supervision. When you need additional help, a colleague at a remote location will see through your eyes to guide you, or just confirm that you performed the task correctly. With predictive analytics fed by data from thousands of other engine repairs, your glasses will identify which parts will likely require servicing in the future, so you can address those issues today. They'll also automatically order replacements for spare parts as you use them.

For companies, the efficiency benefits of seeing rich metadata in context are enormous. Being able to look at any object to trigger a visual search and display relevant data, history, training, and predictions from that object's digital twin cloud datastore saves time and effort. The bigger and more complex the machines and factories, the faster the ROI. Downtime is expensive; with predictive maintenance algorithms, machinery will break down less frequently and come back online more quickly, meaning production lines will run more efficiently and cheaply. And since technicians will have their hands free when working instead of consulting cumbersome screens or paper manuals, safety will improve. On the broadest scale, companies will be able to deploy computer vision to monitor vast industrial infrastructure, dramatically cutting costs, improving service, and enhancing safety.

What will be less appealing to many businesses is the inevitable introduction of AR apps like BMW's from the factory floor into the public market—because with SuperSight, everyone can be an expert. With the proper parts in hand and

An electric SUV "explainer AR" turntable at the Frankfurt auto show.

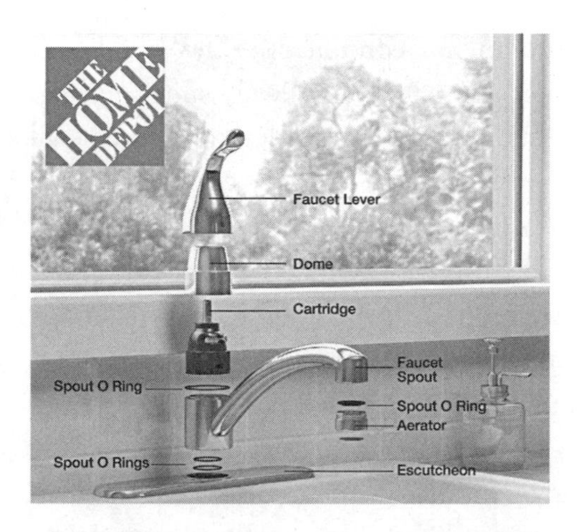

DIY repairs aren't as inscrutable with exploded views and step-by-step guidance.

3D visual guidance that ensures every step is performed correctly, even people who feel mechanically challenged will gain the confidence to swap a plumbing gasket, replace a lawnmower belt, and refill car fluids.

Car manufacturers whose dealerships run on service fees will need to pivot, but other businesses will be pleased to have customers take care of simple maintenance tasks themselves, since customer-support interactions and product returns gouge profit margins. Self-service economics will ultimately make consumers smarter and more empowered, and products more affordable, long lasting, and therefore more sustainable via a circular economy: when we can fix things, we're less likely to throw them away, and more likely to purchase used items we can fix.

SuperSight-enabled guidance will make DIY repairs easier. But it will also help in the workplace when training those new to a specific task or their career.

8.2 Holographic masters to guide you

augmented training simulations

As YouTube and Instructables.com show us, we are well into the age of democratized expertise. Want to learn how to repair a faucet or feed your sourdough starter? You don't need a training course or culinary class; you just need to watch a three-minute video from someone who did.

Spatial computing, with its ability to anchor content onto the world, will drive the next YouTube-scale DIY content company, one ideally suited to help with any problem on the spot. Craft projects, 3D printing, gardening, drones—all hobbies become so much easier when expert guidance is projected directly onto the object of interest and whispered in your ear.

Rookies frequently benefit from the direction of someone with more experience—especially if their task is dangerous or involves expensive materials. For everything from social interaction and sports coaching (as we saw in chapter two), to first responders and farmers, augmented training is AR's killer app.

One huge market here is corporate training. Today this content is typically dry and nearly always delivered ahead of time as a theoretical "just in case" (e.g., just in case there is a polar vortex due to climate change, the electrical grid goes down, and the natural gas lines all freeze, say). Months or years may pass between learning knowledge and putting it to use, which means much of it is forgotten. This results in costly mistakes, wasted material, personal injury, and a terrible reputation for the training itself. (Why else would you be plied with so many mini-muffins during those boring work seminars?) It would be much more efficient if training could be delivered just in time, in context, and only if needed.

One of today's most-used SuperSight training applications was founded by a pro athlete who understands the importance of lifelike context over classrooms and conference rooms. STRIVR is a virtual reality platform founded by Stanford football player Derek Belch. His research found that immersing players in the middle of complex plays helped them better remember those plays on game day and make fewer errors on the field. It was so effective that many NFL teams purchased VR headsets and started creating 360 videos during practices to allow players to re-experience their own perspective as quarterback or tight end or whatever, and move through the action again in slow motion. The idea is that by entraining behavior, coaches can make players' decisions instinctual.

STRIVR also found that what is effective for football plays also works at Walmart—where the playbook to rehearse is less about passes and blocks, and more how to handle irate customers or a Black Friday sales stampede. "VR allows associates to experience a lifelike store environment to experiment, learn and handle difficult situations without the need to recreate disruptive incidents

A machinist creates an experiential instruction manual by simply doing his job, using Expert Capture by PTC.

or disturb the customers' shopping experience," Walmart wrote in a blog post. "Ultimately, everything associates do is geared toward giving customers the best experience. Through VR, associates can see how their actions affect that. It's helpful for associates to see mistakes in a virtual environment and know how to deal with them before they experience it in real life and don't know what to do."

In a brilliant business pivot, STRIVR went from having a market of 1,700 NFL players to serving 2.2 million Walmart workers. And this new form of retail training doesn't require a classroom or a trip to HQ. Employees can dip into a training experience or refresher on best practices whenever store traffic is slow.

STRIVR is also being used to train construction site employees on safety. It's one thing to read a manual about how to operate heavy machinery, or to be told the perils or quirks of a piece of equipment. It's quite another to feel the vibrations of that machinery under your (virtual) feet, or experience safety close calls when you don't abide by the instructions. It's also much easier to mimic the motions of an expert hologram as they work through a set of tasks than to work those actions out yourself from written instructions.

A colleague of mine from MIT, Sara Remsen, also took on improving workplace training for industrial settings, such as manufacturing, construction, mining, and pharmaceutical labs, but she used AR rather than VR. She launched a company that makes authoring and using training modules invisible and natural. Because it's based in AR, her software, called Expert Capture, is ideal for tasks that involve one or more physical tools or locations. If a part must be milled, then polished, then tested, the software projects ribbons through the work environment, like a piece of string to follow, to guide trainees hands-free from place to place.

Being able to train the next generation on how to safely and efficiently perform physical tasks has never been more important. Industrial manufacturers are expected to face millions of unfilled jobs in the next decade as skilled experts retire. (Maybe that's why Sara's company was acquired soon after it was launched by industrial software giant PTC in Boston.) Now is the time to evolve how workers access and learn critical information to reduce waste and training costs, and save manufacturers tens of millions of dollars a year. SuperSight is the linchpin technology for making that happen.

⚠ HAZARD: COGNITIVE CRUTCHES II

Expert guidance via augmented reality has the potential to make so many things easier. But as you consider the human + computer dyad, just how atrophied should we allow the human side of the system to become? Computers already do most of the diagnosis in the auto shop; if they are going to walk mechanics through every aspect of complicated jobs, how soon will *human expertise* become an oxymoron? If expertise is in the cloud, it need not reside in mechanics, pilots, doctors, lawyers, or service techs.

The more automation we embrace, the more we lose the skills and muscle memory to perform without it. In *The Glass Cage*, Nicholas Carr argues that these losses are tragic: autopilot led to a skills atrophy in airline pilots and caused catastrophic crashes; AutoCAD has routed architectural creativity and spawned a "snap to grid" sameness that's easy to observe in the homogeneity of the glass towers filling our cities. We've become so reliant on GPS-powered turn-by-turn directions that our children have lost any ability to read maps and orient themselves in the world. As the Royal Institute of Navigation's former president Roger McKinlay laments in *Nature*: "If we do not cherish them, our natural navigation abilities will deteriorate as we rely ever more on smart devices."

Yet the same "losses" argument might be made for any app today or assistive technology across time, from the cotton gin to calculators. Alarm clocks and watches robbed us of an inherent sense of time. The written word choked off a rich tradition of oral storytelling. Electronic music killed acoustic production. Video killed the radio star. Maybe, but there is also a leveling-up effect that counters those losses. We no longer have the hunting acumen we once did, but grocery stores give us more time to read. Our fast-calculation skills aren't what they used to be, but many more of us know how to program by learning Scratch in elementary school, the basics of robotics in middle school, and Python by high school. You can't re-land a rocket on a moving drone ship with fast math skills and a good sense of direction alone.

One of the areas in which I'm most concerned about automation and machine learning is those pastimes that give us pleasure because of their uncertainty, and because of the work we put in over many hours to develop our hard-earned skills. Tools like fish-finders that show precisely where to cast with which lure and at which depth, or augmented ski goggles for snowboarding that paint a very narrow line for you to carve down the mountain, are powerful tools that, while offering shortcuts to fishing and to mastering steeper slopes, feel like they rob us of exactly the reason we fall in love with fishing and free-riding in the first place. When precise guidance puts guardrails on the unbounded panoramic expanse of the ocean or mountain, you may be creating blinders that limit your experience of adventure.

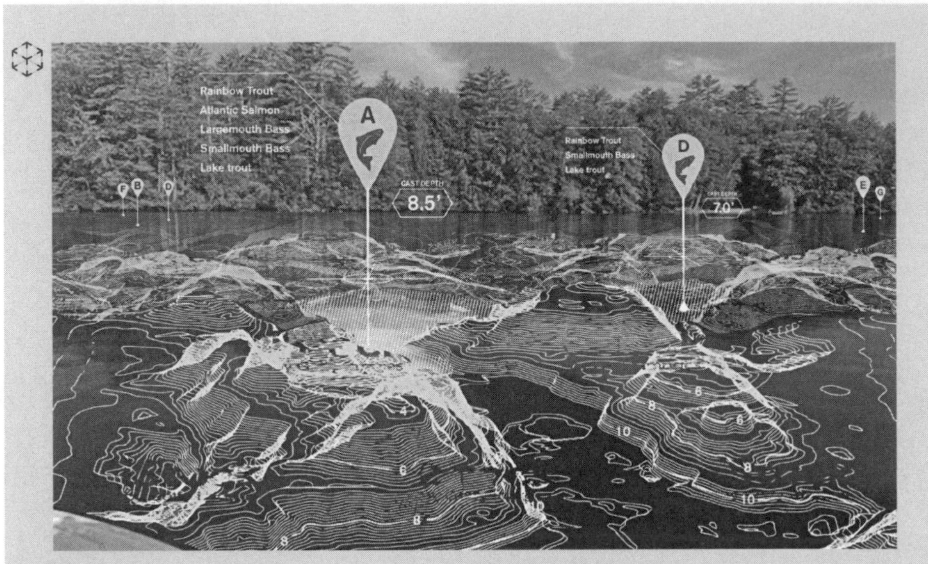

CleAR Water superimposes underwater topography on a lake, with machine-learned probability clouds for catching certain fish at precise depths with particular lures. Is this the natural evolution of electronic fish-finding? Or is it too assistive, making time on the water overly prescribed and removing the opportunity to develop learned instincts—not to mention eliminating the essential randomness and boredom that are a part of a good fishing excursion?

8.3 Leaving your mark on the world

history of hobo marks *spatialized messaging*

One place in which we urgently need to embed and anchor expertise is our aging infrastructure: electric grids, waterworks, bridges, and transportation networks. A graying population knows how these complex systems and machines were built, their particular quirks, and how to keep them performing. But what is second nature to someone with decades of heavy machinery experience is a strange land and foreign tongue to the digital natives coming into these jobs. They are accustomed to random swiping and tapping, not following an intricate sequence of steps with manual levers and analog regulators. They don't want to go old-school Space Shuttle; they expect SpaceX.

This presents a significant challenge: How do we best pass on older workers' hard-won understanding of aging vital infrastructure to a younger generation of workers? Instead of dog-eared, yellowing paper manuals to train younger generations, might we be able to leave useful labels, marks, and annotations directly on those old machines and their surroundings?

The technology that will allow for this kind of cross-generational visual integration doesn't come from the future, but from a fascinating story in the past.

Following the Great Depression in the 1930s, millions of people in the United States lost their jobs, farms, and, in many cases, homes. These vagabonds developed an asynchronous language to communicate and help each other out: marks on fence posts, walls, and sides of homes, called "hobo marks," that can still be seen today.

The language is made up of a series of hieroglyphics, each designating things like whether a home was inhabited by a kind-hearted lady or a dishonest man, whether it was an unsafe place or one where you might safely spend the night—or even if there was work for food. A triangle facing down meant that other hobos were there, and a triangle facing up with a little tall hat showed that the homeowners were wealthy. All of these marks helped homeless people find doctors, places to sleep, or their next meal. It was a way of crowdsourcing support and literally rendering knowledge on the landscape, all with a pocketknife on a fence post.

What's wonderful about hobo marks is that they're a persistent, iconic, encoded, and context-specific way to codify information. It's a *spatialized knowledge sharing system* literally marked onto the landscape. And now, AR technology lets us leave our own digital hobo marks on the world for others to see.

AR hobo marks drawn over restaurants on the island of Nantucket. These spatial notations helped other conference-goers find the best places to go or those to avoid.

I ran a workshop at a conference for innovation leaders in summer 2019 where I gave everybody an assignment. First, I had them install World Brush, an app through which people make marks on the world around them that are visible through their phone camera. The marks are anchored in place and shared, so others with the app who wander by can see them, as well as a 2,000-foot-view map of all marks in the area. Then, I asked workshop participants to start marking up the world with persistent AR marks. We were on Nantucket, where the downtown is only six blocks across—dense enough that the workshop participants were likely to bump into each other's marks over the course of a few days.

Everybody went out and started painting symbols to help each other; there were lots of marks for good restaurants, for example. It made me wonder: Instead of asking your friend for a list of places to eat and things to do when you come to visit, what if you pulled up World Brush and could see their recommendations as stars over physical restaurants, shops, and other places around town?

The process of working with tools like World Brush always brings up a lot of interesting design and data policy questions. For instance, how long should these marks persist? The workshop was one afternoon, and people were going to stay on the island for just another day or two after that. Would they want those same marks to persist into the next summer? Another question is how many other people should see these marks. Our workshop of twenty people obviously wanted to see each other's marks, but should the same ratings and marks be available to other visitors to Nantucket? Or would they only confuse and frustrate the locals who are living there the other 360 days of the year? When is a designation helpful, and when is it just virtual graffiti?

Neuroscientist Beau Lotto designed a location-based messaging system called Traces, which illuminates the design questions at stake when spatially anchoring messages on the world. Rather than sending messages to other people regardless of where they are, Traces lets you leave notes in specific places—text, audio, visual, whatever you want—for others to see only when they arrive there. You choose how long the message will stay there (an hour? a day? a year?), and the scope of who can see it (everyone who works at a company, or the whole world). When someone within the permission scope arrives at that location, they receive the message. For example, I might leave a Trace outside a lecture hall to introduce a guest speaker to latecomers as they arrive, or drop a recommendation for

THE DURABILITY OF VIRTUAL MARKS IN THE WORLD

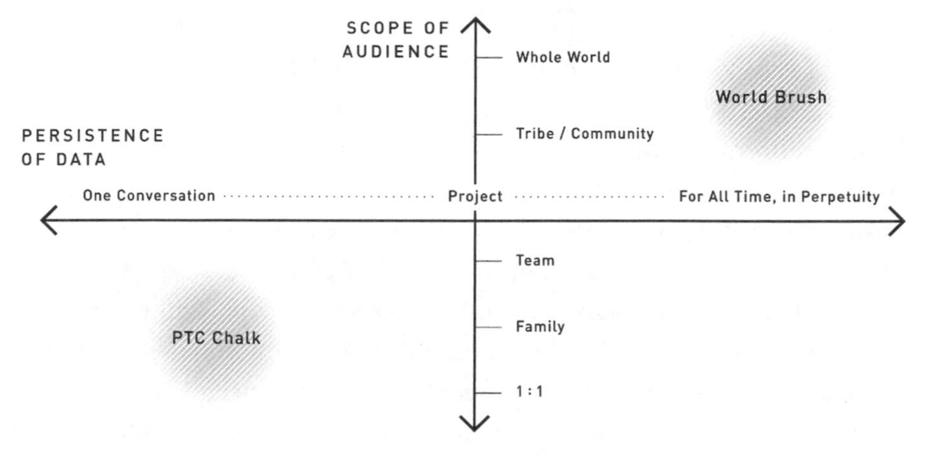

This framework shows the vast lifespan of virtual marks, from those that only exist for a few seconds to those that remain forever, and the scope of who has access to these marks, from just one other person or the world.

the tiramisu outside an Italian restaurant for any friends who eat there. Leaving hidden notes around the world has a certain spy-culture romance to it!

Helpful notations on the world would be relevant for many kinds of work. Instructions for use could be left on equipment or tools, or checklists and repair logs could be attached. Some of these virtual marks should stay in place permanently, like DANGER HIGH VOLTAGE. Others, like notes to oneself, might persist for the duration of a service call; once you depart, they would disappear. Being able to specify the audience who sees each mark would be helpful, too—perhaps junior power plant employees see more detailed instructions than experienced staff. Anyone who has reset the electrical grid incorrectly in the past gets a little extra help.

SuperSight's potential for capturing and embedding expertise about how to use things, then delivering that knowledge at the point of action where it's needed most, tailored by level of need, will demystify millions of situations. But perhaps an even greater potential workplace application for SuperSight is improving safety.

8.4 Predicting accidents before they happen

job-site safety *pooled training data*

Construction is one of the most dangerous jobs on the planet, according to the Census of Fatal Occupational Injuries. Every day, builders are required to work from heights on ladders and scaffolding, operate heavy machinery and power tools, and perform other hazardous tasks. In this physically demanding line of work, injuries are common and fatalities a reality. How might we use SuperSight to make these environments not only safer but more effective?

Construction is the perfect Petri dish for SuperSight because the job is so physical and, as we saw in chapter six, involves so many moving parts. While there are always detailed digital drawings and building information models for completed structures, actual job sites are a multidimensional hot mess.

"Planning is the only calm part of this business," Jit Kee Chin confessed. "It's controlled chaos after this, as hundreds of unpredictable things go sideways." Jit Kee was brought into Suffolk after a decade as a strategy consultant for McKinsey, and she is one of the sharpest analytic minds I've ever met, as well as a standout in the male-dominated world of construction management. As the company's

Smartvid, a SuperSight platform to help keep construction sites safer by spotting people without proper safety gear.

EVP and chief data and innovation officer, she has methodically analyzed the root causes of delays and risks in Suffolk's many building projects around the globe, and has made a set of venture investments in promising companies that solve these issues.

"Safety is a huge challenge for construction," Jit Kee said from the stage at MIT's EmTech Conference in 2019. To address this need, she invested in Smartvid, a SuperSight-powered startup that aims to keep job sites safer by recognizing many types of risks, and worked with them to develop algorithms to predict injury based on subtle visual signals.

When crews don't wear hard hats consistently, work on upper stories without clipping into safety systems, or pass too close to dangerous machinery, these events are captured and logged by Smartvid.io's system. Image feeds come from many sources: from the photos taken on mobile phones to track progress, fixed cameras, or wearable cameras placed on workers' hard hats. When the system spots indicators of safety risk, including Occupational Safety and Health Administration (OSHA) violations, like a ladder being used improperly or precariously piled-up building materials, these events are flagged, compiled into dashboards for workers and managers to review, and then fed into algorithms that convert the behavior into actual risk metrics and likelihood of injury. Think of it as gamified safety with a smart feedback loop.

This deep-learning system is a massive probability calculator that estimates the chance for future injury and delay on that site. Predictive analytics can then recommend a prioritized list of actions to take on each job site. According to OSHA, eliminating the "fatal four"—falls, being struck by an object, electrocutions, and being caught in between equipment—would save 591 workers' lives in America every year.

Suffolk and Smartvid knew they needed more data to feed and improve their algorithms, and that job safety should be a shared service for the construction industry, not a proprietary advance just for Suffolk. So they made the product available to competitors, and created a consortium in which companies using the predictive system all contribute a wide range of data to improve its performance. By sharing data and best practices, the entire industry learns faster, and organizations benefit from both improved system performance and the ability to compare their construction projects against industry benchmarks.

This consortium approach is needed especially for deep-learning systems because more shared data means more precision and a more robust product for all members. Through this consortium Smartvid learned that multiple data types, including images, project data, payroll data, incident records, and more should be combined to most accurately predict risk. The combination of construction site images with other project-related data enables the system to become better at detecting hazards and assessing how much risk they actually represent. Due to the power of this novel combination of SuperSight (or "unstructured") data with "structured" project data for predictive analytics, Smartvid was relaunched as Newmetrix in August 2021.

While it may seem counterintuitive to want to share one's safety information with competitors, many companies wouldn't have enough data on their own for their Smartvid systems to make good decisions. Sharing data means improving safety, and that's an incentive no construction company would turn down.

This kind of SuperSight predictive technology will be used to increase safety outside of construction sites, too, wherever insurance companies cover accidents. In 2018, I keynoted a conference for AppFolio, a company that makes software for property managers. They're interested in applying computer vision to flag safety hazards at the thousands of residential properties their customers manage. If a security camera can spot black ice, toys, or beer bottles left on stairs, or trash that needs picking up, AppFolio could ensure a safer environment for renters.

More generally, mobile eyes in the form of car dashboard cameras and drones will automatically spot driving hazards like downed trees or power lines, objects on the road, or large puddles after a rain deluge. Citywide, these cameras will calculate risk on behalf of 311 services, like potholes or cars parked in bike lanes, then recommend improvements supported by a probabilistic model: change this intersection for $130,000, and you avert twelve accidents over the next decade and save $360,000 in hospital bills and claims-processing costs. Given the findings of this hypothetical city-safety clairvoyance service, insurance companies and self-insured employers will be happy to fund a safer city to avoid these costs.

As we increasingly use SuperSight to avert injury for dangerous jobs, and optimize yield in endeavors from manufacturing to fishing, might there be a backlash to a precision-first future? I believe we are already seeing a countervailing trend in reaction against this modern Le Corbusier steel-and-glass perfection—one that rejects hyper-quantified precision in favor of craft, the charm of minor flaws, and the beauty of asymmetry.

8.5 Manufacturing authenticity

problems with digital perfection *delights of imperfection*

One of my cousins (you know who you are) obsessively scours art fairs and ceramic studios to find mugs, bowls, pitchers, and plates that, to me, look like they were made by a five-year-old. The saggier the handles, the more splattered the glaze, the better. Precisely because there are entire stores filled with pristine, machine-washable tableware, she prefers the mugs and plates with the most flaws (she would say "character"). Her dinner guests immediately recognize that every one of them is bespoke and handmade—and costs $80 versus $2 at Target.

It's now a rarity and privilege to have something that looks like it was made badly. That's because it's cheaper to make 5,000 identical things via automation than five variable things by hand. Imperfections in clothing, the patina on a brass door handle, vintage worn leather furniture, even the ragged paper edge of your "homemade" holiday card, are signals of heritage and privilege.

It's not even just a little variation that we crave—we're beginning to prefer things that have *a lot* of skew. Factory automation and the machine age gave us consistency and affordability, but also a boring sameness. These days, we want to see signs of the human in the loop.

To make a classic Danish-design Hans Wegner chair, a craftsperson shapes the legs on a lathe, carves and sands the back and arms by hand, and executes the dovetailing and joinery work with wooden pins and glue before weaving the

A classic Hans Wegner chair features complex curves, executed with minor variations that could only be hand hewn. Can computer vision guide and teach this kind of craft?

rattan seat in place. There are considerable allowances for variation as long as the final chair is balanced and symmetrical. If you look at a set of these on a collector site like FirstDibs.com, you will notice that each chair is beautifully, slightly different—and costs over $5,600.

Compare this craftwork to what SuperSight-enabled companies like Cognex, Industrial ML, and Vision Machine bring to robotic manufacturing. These companies use computer vision to detect minute defects in products as they are manufactured to dramatically boost quality, reduce waste, and optimize company profits. Perfect replication is the only shippable output; there is zero tolerance for error. This kind of precision is essential in auto-fill pharmacies, where you need to populate small bottles with, say, exactly thirty pills. These pharmacies use SuperSight cameras to count visually and in less than a second, too fast for the human eye.

There are clearly productivity advantages to these technologies, and for industries where proper function depends on getting the details exactly right every time, there are safety advantages, too. Every kitchen appliance you buy, every plane you fly, and every car you (don't) drive will soon be flawless thanks to cameras and computer vision systems that manufacture perfection. These kinds of businesses want more eyes on the line (literally) to scrutinize and survey things being made and create them with more precision. And in general, automated scrutiny narrows tolerances, reduces waste, and decreases downtime.

But these cameras are also squeezing any sense of hand-built craft out of objects and the environment, with negative consequences for our relationship with and attitudes toward the objects we use every day. We may start to regard everything around us as commodity, not craft—as expendable clones to use and discard rather than things to treasure and repair. And the more perfect our manufactured items become, the more we crave and value authenticity, originality,

These IKEA chairs only look delightfully handcrafted.

variation—evidence of the maker's hand. That lopsided, hand-thrown, clay teakettle may not keep your beverages as warm as a double-walled aluminum thermos—but it's beautiful.

In an ironic twist, some companies are now using computer vision to manufacture imperfection, "designing in" seemingly authentic blemishes. IKEA, for example, launched a line of furniture and housewares from the famous Dutch designer Piet Hein Eek, who is known for showing the hand of the maker in his work. The bowls aren't round, the chairs are slightly askew, and they feel more authentically crafted and imbued with charm and personality as a result. "You can feel a personality behind each of the products and know that each is unique," IKEA's creative leader, Karin Gustavsson, told me when I was at the company headquarters in 2017 to present at their yearly Democratic Design Days conference. The line's ceramic vases have lumpy forms created from dozens of different handcrafted molds. Its glassware was created by two different machines, giving the full collection a mismatched feel. The irony here is, the same mass-production techniques were used—the imperfections were just captured in the mold.

This "flawed charm" is evident in nearly every expressive medium. In music, for example, we appreciate imperfection: the sound of hands moving across guitar frets, drums not quite in time, the crackle and hum of an LP record. Though some producers still rely heavily on Auto-Tune, many will "un-quantize" musical recordings, adding an expressive tempo rubato, so that electronica sounds warmer. Otherwise the track can seem too synthesized, robotic, and unnatural to our ears.

With so many products we buy, it's the imperfections we love, crave, and will continue to pay for. If you're a dressmaker or a potter, computers might well be able to do your job more accurately than you—but for consumers who still want that human touch, even if it means mistakes, that spells opportunity. Bespoke items and makers are, in a sense, automation-proof—that is, until computers perfect the ability to manufacture imperfection, too.

What this demonstrates is that, within the coming world of SuperSight-enabled augmentation and automation, there is still a very real and urgent role for human sensibility—strongly supported and assisted by incredible new tools. Nowhere is this more true than in the way we approach imagining and communicating what's next.

What if we could see the future?

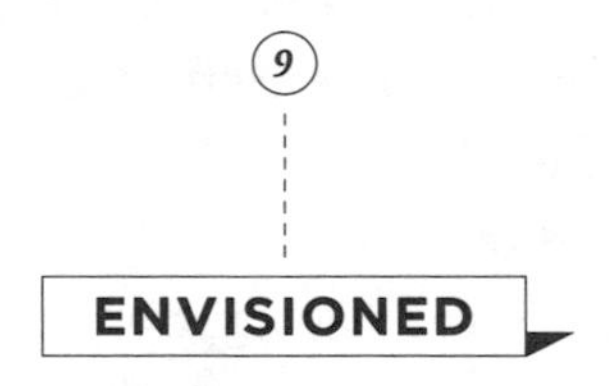

About visualizations so compelling they inspire change

9.0 FutureCasting

evolutionary biology *expanding our perceptual envelope*

It's hard to see the future—even for a futurist. We use tools like scenario planning, prediction markets, and design fiction to explore possible futures. Then, like those World War II radar operators in Dover from chapter one, we look for weak signals that might grow into important trends. We try to connect the dots from today's lab inventions to tomorrow's big commercial opportunity. How fast might a new technology reach the tipping point of viability? Often, this is best done with log-scale graphs, since many technologies advance exponentially. Given trends in battery density, a straight linear extrapolation wouldn't have predicted the opportunity for electric cars (or even electric planes), but an exponential lens makes battery-powered buses, trains, boats, and aircraft inevitable. The same analysis of smart glasses reveals a lot of key components advancing in exponentials: hardware miniaturization, memory costs, compute power, display density and dynamic range, and the AI algorithm performance that presaged SuperSight.

While trying to see the opportunity spaces that open as a result of the combinations of these technologies, we scan for secondary and tertiary effects, like what happens to gas stations when we charge at home, and how companies that rely on service station sales might respond to this shift. Rather than simply observing and measuring these shifts and sounding dystopic alarm bells, I prefer to envision and prototype desirable future scenarios. Might filling stations pivot to serve another convenience need, with their drive-through placements in the landscape and giant storage tanks below? (Maybe hydroponic fertilizer distribution for your agriculture-at-home efforts?)

Large-wave macro trends like digital disruption, cloud computing, mobile and voice interfaces, Internet of Things, and VR are easy to anticipate; correctly timing their wide-scale adoption and absorption into culture is harder. Even hazier are the myriad permutations of products, services, and business models that startups and entrenched brands will launch. The exploding complexity of these industry interactions makes the job of a futurist akin to a game of multidimensional chess. It's nearly impossible to foresee how technology waves will play out, and even harder to divine their timing. For instance, few planned for a global pandemic, its profound effects on the restaurant and travel industries, and how it accelerated adoption of ecommerce, telemedicine, distance learning, home workouts, and remote collaboration.

As our geopolitics are in flux and our climate is reaching a tipping point, we must engage with urgency with the discipline of foresight. Thankfully, SuperSight provides the tools needed to fathom and plan for these shifts, and that is the subject of this last chapter: using augmented reality to imagine, invent, persuade, and secure a better future for us all.

SuperSight, I'd argue, is best put to use as a foresight tool. And the explanation for why starts a long way back on our evolutionary tree.

In the dark, murky rivers of the Amazon, there is a fish without eyes. The black ghost knifefish hunts by emitting a weak electric field that extends a few centimeters away from its body. When this sensory field is disturbed by a poor little Amazonian tadpole swimming into its path, the fish lunges for its lunch. The space and time between stimulus and action is very short; being able to think ahead would confer no particular advantage here.

Now contrast the way a lioness stalks her prey. She selects the most delectable-looking antelope she can spy on the savanna—organic, free-range, the good stuff—then edges along a line of bushes to conceal herself, carefully choosing the timing for her attack.

What motivated the evolution of the lion's shrewd planning brain, versus the ghost fish's reactive brain? Surprise, surprise—it's the eyes.

I was so delighted about this evolutionary link between visual performance in the animal kingdom and complex cognitive tasks like planning that I visited

"In what ways do we need our eyes to further evolve today, to help humans survive and thrive?"

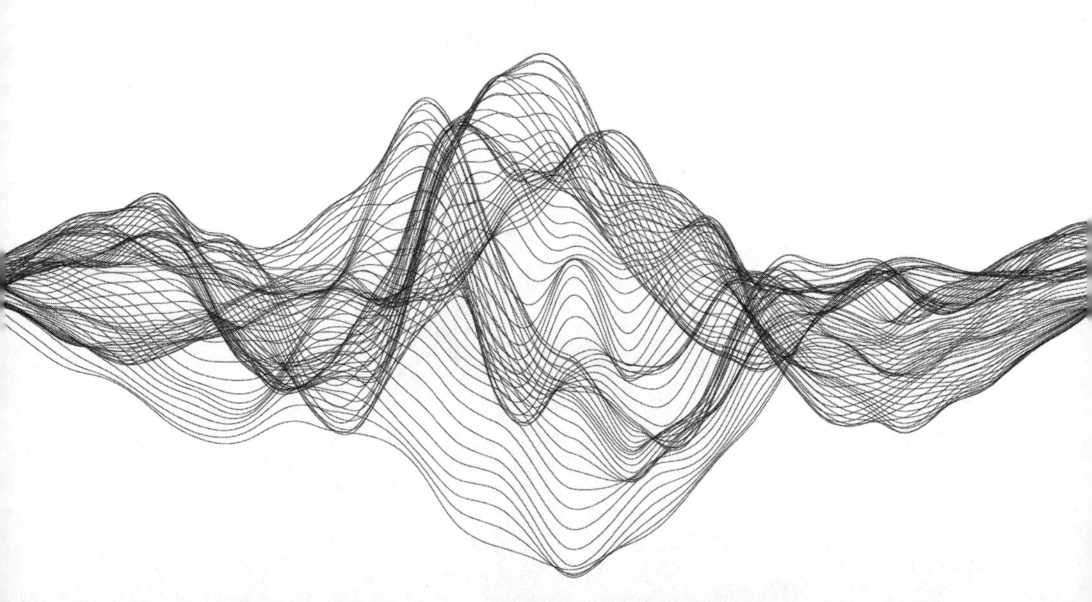

neurologist Malcolm MacIver—who named this "the buena vista hypothesis" and wrote a landmark paper describing it—in his office at Northwestern. There, he showed me a chart explaining the evolution of sight in our predecessors, and what he calls their *perceptual envelopes.* The chart went from single-cell autonomous animals that could see about one ten-millionth of a meter in front of themselves, to a peregrine falcon, which, on a clear day, can see a rabbit from two and a half kilometers away.

Fossil evidence shows us that we come from myopic, small-brained vertebrates much like the black ghost knifefish in the Amazon, which then began to peer above the surface of the water and see farther. Fifty million years later (these things take time), our ancestor the *Tiktaalik* started to lurch up on land to eat shoreline centipedes. The reason? They could see them; their perceptual envelope had become much larger. Light attenuates and dims rapidly in water; in air light propagates much farther. (This is why submarines and sea mammals use sound to communicate, not visual signals or fin gestures.) This massive increase in visual range—the ability to survey the world above water—triggered a gigantic adaptive effect on animal cognition; the brain grew a thousandfold and evolved the now-useful ability to *plan.* It was sight, not legs, that brought our ancestors out of the water. We went from surviving in a purely reactive mode to needing to reason about the future.

In the water world, with a small perceptual envelope, strategizing and scenario planning weren't useful. But in the air world, better eyes meant we could see predators and prey through trees or across a pond—and they could see us, too. For the first time in our evolutionary history, it really paid off to contemplate different scenarios.

Tiktaalik, our evolutionary ancestor from 350 million years ago, peers out of the water and suddenly needs a planning brain and some legs to transition out of the water and catch distant prey.

So, better sight gives rise to planning, which is in turn the genesis of consciousness. Consciousness, as psychologist Bruce Bridgeman puts it, "is the operation of a plan executing mechanism, enabling behavior to be driven by plans, rather than immediate environmental contingencies." Or as MacIver summarized, "Thinking ahead is not worth doing unless you can see ahead."

My central question for MacIver that afternoon in Chicago was this: In what ways do we need our eyes to further evolve today to help humans survive and thrive? Do we need to see farther like the falcon, or in the dark like an owl, or into the infrared spectrum like a hummingbird? Given the vast capabilities of SuperSight glasses to mix any information into our vision, what can we do with it to radically improve our vision? *How do human eyes need to evolve?*

In modern times, seeing farther doesn't provide an evolutionary fitness advantage. (Not being able to see fast-food restaurants, on the other hand, might.) I'm more likely to need +2 diopter reading glasses to discern on a tiny food label if there are allergens in those crackers than I am to need a telescope to spot a rabbit from a long distance or in darkness. Being able to see far into the distance on the savanna helped us plan better for hunts, but it's not exactly a survival mechanism anymore.

What if we began to think about visual acuity not in terms of how far we can see into the distance, but how far we can see into the future? Our pressing modern problems are things like climate change, racism, social injustice, poor access to healthcare, food and water scarcity, and (ironically) the existential threat of AI stealing jobs. Humans need tools to understand and tackle these issues. This is the direction in which we must evolve: to see farther into the future, imagine alternatives, and weigh the important ramifications of each. I call this discipline *FutureCasting*.

Human beings tend to be terrible at making decisions with long time horizons. We struggle to comprehend and reason about the distant consequences of our actions. (Anybody who's been watching what they eat but has given in to the 10 PM temptation of mint chocolate chip ice cream knows what I'm talking about.)

Behavioral economists call this temporal nearsightedness *hyperbolic discounting*: valuing rewards we receive immediately, today, over larger rewards we receive later. What is happening *right now* feels so important that we tend to ignore or discount the future. We don't save enough. We overcommit our time and attention to far-off events.

So, what we might need most from our evolved eyes is metaphorical foresight. With prepared food available on every corner, we may not need to improve our visual acuity to the level of an eagle to find our next meal. However, our wellbeing as a species would benefit from the ability to vividly "see" the distant consequences of our actions.

We've seen opportunities over the last chapters for SuperSight to assist us across time scales: enriching today's interactions with others, making near-term plans like this week's cooking and shopping, and setting medium-term goals like learning new skills. But in this chapter, I want to focus on time horizons five, ten, or twenty years out. SuperSight has the greatest potential to help us vividly imagine and bring these futures closer by communicating them with veracity and in context, and in the process motivating action and commitment towards the future we *want* to manifest. For example, what problems require a bird's-eye view to analyze and imagine? What solutions might we be able to see at a higher altitude?

9.1 Clairvoyance from the sky

satellites with machine learning

In Norse mythology, the ravens Huginn and Muninn sit on the shoulders of Odin, the god of wisdom and war, serving as his twin spies. As one medieval historian describes, they "whisper all the news which they see and hear into his ear . . . He sends them out in the morning to fly around the whole world, and by breakfast, they are back again." This gives Odin a certain omniscience for observation and discovery, and thus earned him the name "raven-god" (*hrafnaguð*).

The myth of Odin has come to pass for anyone with the means. We now have thousands of ravens to see for us. They spend the day circling the earth, watching what is happening to the people, weather, buildings, and boundaries of our nations in minute detail, and reporting back everything we need to know.

And they're not the stuff of legend—they're the stuff of science.

During the Cold War, only a few nation-states had the resources to deploy planes and satellites to gather surveillance images, much less the staff of human analysts needed to decipher salient details in the deluge of visual data. Today, microsatellites, high-resolution digital photos, inexpensive cloud storage, and computer vision combine to grant anyone powers of observation that would make a Norse god weep.

Odin was given omniscience from above through the eyes of ravens who circle the earth daily—or so the story goes.

Eyes in the sky change the way we gather information. Vast streams of visual data come at us from above: one company, Planet.com, has more than 175 satellites, providing daily coverage of the entire planet at sub-meter resolution. Add algorithms that spot patterns in the data, and we're entering an age in which it will be virtually impossible for countries, companies, or anyone else to keep secrets—in which everything physical becomes transparent.

SuperSight is ultimately a planetary-scale phenomenon. Think of how radically our consciousness of ourselves and our planet changed when humanity first saw the pictures of Earth taken from orbit around the moon, a photo commonly referred to as *Earthrise*. Today we can view our world as a continuous canvas, stitched together from an increasingly detailed patchwork of satellite imagery, just by opening Google Earth. Governments, organizations, and individuals can now survey anything in the world from above, without needing a plane, drone, or ravens.

Expanding our machine-readable gaze to the planetary scale—and asking planetary-scale questions—becomes easier as the price of this data falls, pay-per-use business models lower transaction costs, and our automated search improves. Using a neural network trained for similarity searching, we can Google anything visual on the planet, combing through petabytes of data in just a few seconds. The pattern-matching magic we first experienced in audio with Shazam—where you hold your phone up in a bar to sample a few seconds of a

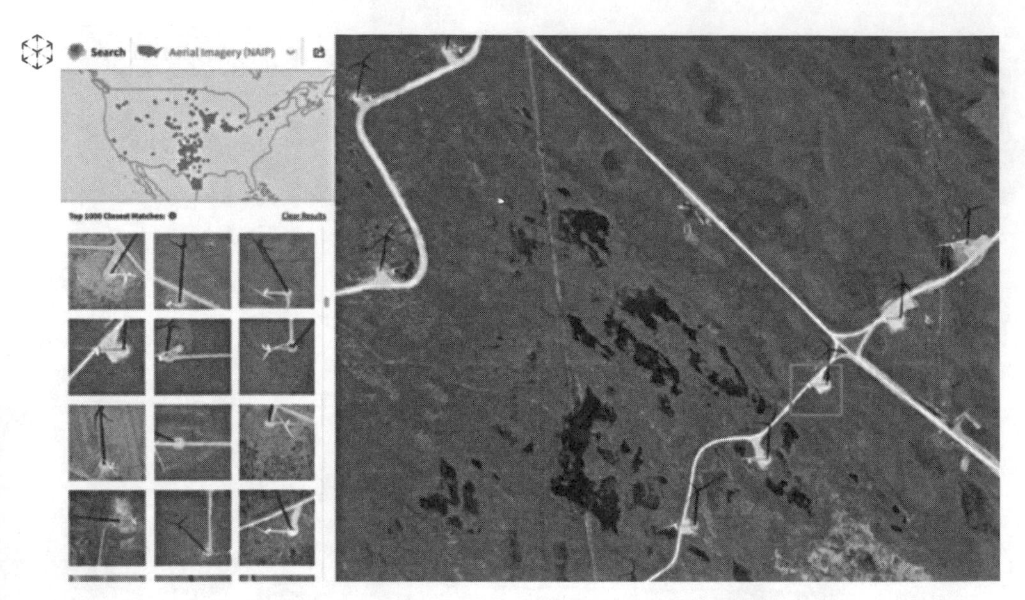

I chose an image from a wind turbine, and the matching algorithm from satellite data company Descartes Labs found similar images from around the globe in less than a second.

song to match it, save it, and sing along with the lyrics—has now come to the visual domain.

Drop a photo of a pool, a water tower, or train intersection into a similarity search, and the system will find approximate visual matches. The countless uses of this surveillance data, coupled with automated search algorithms, boggle the mind. Say you're a solar installer; which roofs have good sunlight but no solar panels? Computer vision algorithms can now answer this question and perform other tasks that would require a full-time team of hundreds, but in a fraction of the time. This means that industry players can obtain vital information to make informed business decisions more quickly (always) and more accurately (often) than people. Computer vision–enabled satellites also provide more public knowledge of corporate behavior, climate effects, military mobilizations, and humanitarian efforts. Ideally this shared oversight and vivid evidence will bring humanity together to tackle some of the foremost issues plaguing our planet.

For example, Malaysian palm oil companies have long struggled with monitoring the industry's often-opaque supply chain to verify their contract growers are practicing sustainable agriculture. Now, Descartes Labs has provided them the tools they need. Satellite data and computer vision can find deforested spots overnight, which pressures palm oil conglomerates to live up to commitments made in sustainability reports and verify that their suppliers adhere to ethical growing standards.

For satellite data company Orbital Insights, which boasts "transparency across industries," selling data to financial service companies, among others, has been a goldmine. They famously use computer vision to count the number of cars in the parking lots of retailers like Home Depot or Walmart as a proxy for consumer demand. During the COVID-19 outbreak, they monitored production slowdowns for factories by observing employee parking lots and the movement of shipping containers on land and sea. Farmers and commodity traders likewise no longer need to wonder how much supply exists in their industry at any given moment: Orbital Insight's algorithms estimate these important metrics by analyzing aerial images of agricultural fields.

Many types of economic activity are now visible from the sky to investors, competitors, and other interested parties—from occupancy rates at hotel chains, to new home building or renovation, to weekend recreational boating, to the size of crowds at soccer and baseball games. Any advantage that interested parties might have obtained by keeping this information secret is now eroded.

Satellites don't only serve companies; they're also a critical research tool for academics and humanitarian aid workers. Nonprofits and government agencies can assess the size of refugee camps and the flux of migrants escaping war-torn countries. Geologists search these images for mining opportunities, and scientists use them to measure our changing planet.

At Ditto Labs, our analysis of the millions of social media images posted every day helped researchers study country-scale public health questions with computer vision, like the incidence of smoking across many US cities. Now, in the time of pandemic, we have more urgent population-level health data to visually tally: who wears a mask (by gender and age), who is distancing, whose body temperature is elevated. Once we're able to take our masks off, I expect we will do more measurement around population-level interactions and mood, combining social media fire hoses with urban traffic cameras and applying the emotion-detecting algorithms we discussed in chapter two: What cities, and which cohorts, are now socializing? Which are still depressed?

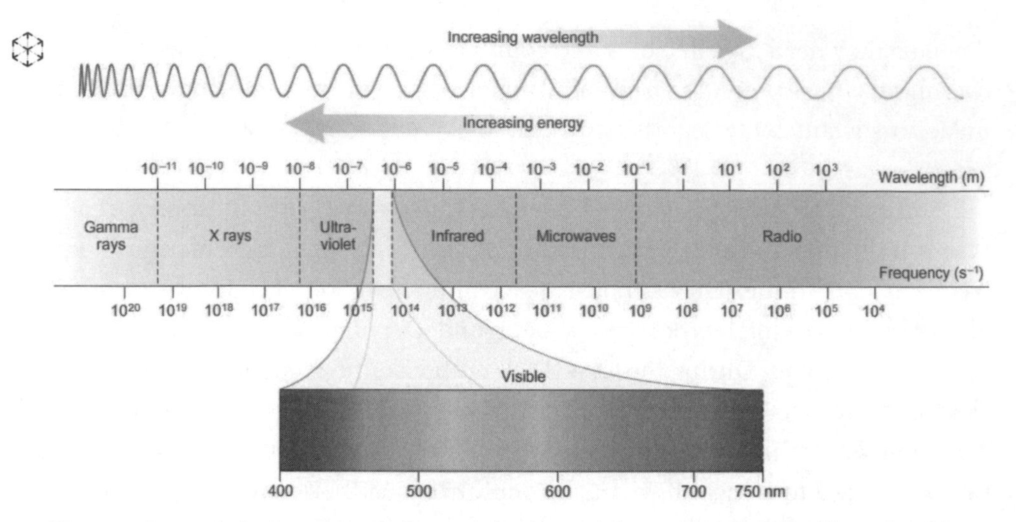

The spectrum of electromagnetic frequencies is much larger than the thin slice of visible light that humans can comprehend. New low-cost sensors onboard satellites, and soon built into smart glasses, give us the ability to see heat, through walls and bodies, and more.

Many animals have evolved to see more than is visible to us humans. I mentioned how birds of prey see trace trails of urine in the UV spectrum. Butterflies, reindeer, bees, scorpions, and salmon have the same ability to see shorter-wavelength light, which they use to hunt or mate. Bloodsucking insects like bedbugs and mosquitoes use infrared vision to locate hosts via their body heat; and many nocturnal animals, like pythons, boas, and rattlesnakes, have highly attuned sensors to detect the heat of their prey in the dark. Modern eyes-in-the-sky satellites are similarly equipped with broad-spectrum sensors that observe wavelengths of light outside human visibility. This extends our perception—for example, through the tree canopy to the remains of ancient cities hidden below.

In chapter five, I explained how 3D spatially anchored projections can reanimate historical sites like ruins and castles. SuperSight in the sky is now helping us discover them, too. What used to take expeditions of archaeologists painstakingly chopping through overgrown rainforests and risking curses and booby traps from pissed-off demons and their worshippers (well, in the movies, at least) can now be done using a plane equipped with LIDAR. These cameras pulse laser light at a frequency that "sees" through the tree canopy and bounces off the ground below, then time the return of each pulse. If you know the precise

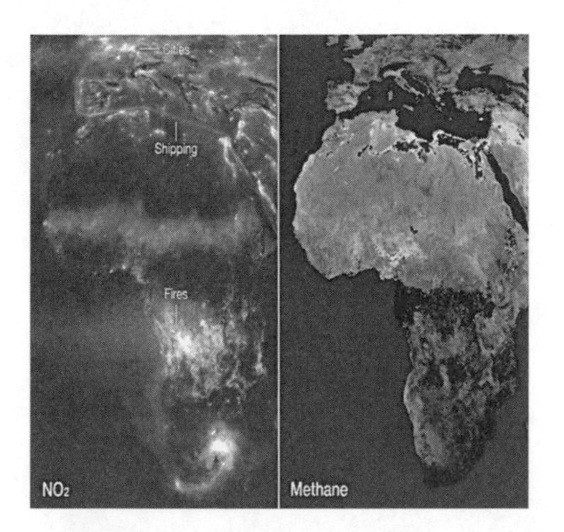

Satellite imagery renders NO_2 and methane emissions in visual heat maps. This surveillance can catalyze conversations about accountability, policy, and how to extinguish global warming.

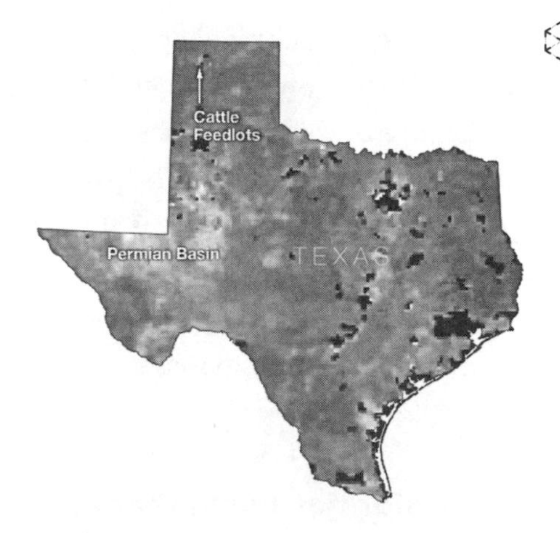

Beef cattle feedlots registering clearly in a January to April 2019 methane composite of Texas.

time and location of the plane when it sent and received each pulse, you can mathematically reconstruct the 3D terrain. At ground level, a field looks like a field and a jungle looks like a jungle. But to the LIDAR cameras, the result is an incredibly detailed and accurate map of a region's topology—including traces of structures that once stood there. Thanks to this powerful new tool, archaeologists are discovering the remains of millennia-old stone circles in England, ancient Maya ceremonial burial sites, and a whole new city near Angkor Wat.

Recently, satellite instruments have been used to analyze earthshine—that is, how sunlight scatters in the atmosphere and is reflected by the surface of the earth. The spectra this generates reveals the concentration and distribution of gases critical to pollution and global warming. In October 2017, the European Space Agency launched the Sentinel-5P satellite, ushering in a new era of atmospheric monitoring from space. Every day, the TROPOMI (TROPOspheric Monitoring Instrument) provides global measurements of ozone, NO_2, SO_2, formaldehyde (an indicator of hydrocarbons), and sulfur dioxide (from industrial pollution and volcanic activity), plus methane, a substantial component of global greenhouse gas emissions.

Looking at NO_2 emissions across Europe and Africa reveals a map that clearly identifies areas where things are burning: Mediterranean shipping lanes, brush fires in Africa, and highly populated cities across both continents. And yes, you can even "see" cow belches and flatulence in the methane emissions spectra.

Seeing these trace gases is not only a new form of omniscience; it's also clairvoyance, as it predicts the pace at which our planet is warming up. This daily planetary-scale surveillance is the record and ground-truth evidence of our possible demise. It shows specific sources of pollution and flags who is responsible for these ongoing offenses—and then, with this Odin-like knowledge, we can use that information to make change.

9.2 A vision for humanitarian aid and wildlife management

transparency and accountability

Originally, I had called this final chapter "Accountable," because one of the most important uses of SuperSight, especially through satellite images, is the power to make humans—governments, nonprofits, and individuals alike—accountable for their actions or inaction. In 2018, Americans gave more than $400 billion to charities, even though the connection between donation and impact is often opaque, or even suspect. What if there were a way for us to see with our own eyes that money was going to the right place? SuperSight has the power to grow philanthropic giving by orders of magnitude, because it powers a better and more specific feedback loop.

The same satellites that can hunt mineral deposits can also provide evidence that humanitarian aid is being deployed as expected—and making an impact. For example, if you're contributing to a project that is building wells for people that need clean water, you want to be sure that corrupt local government officials aren't siphoning funds off the top and reneging on their promises. Charities currently try to provide this accountability in a very labor-intensive, hands-on way: they log reports and photos of what they've done and send them back to HQ. But if you're building any form of infrastructure, such as irrigation or bridges, that same accountability is available more readily by looking down from space.

This is useful beyond preventing fraud. If you can transparently share with donors visual evidence of a humanitarian aid endeavor making progress, it's easier to attract further funding. Charities could offer a time-lapse of the work being done on the ground as part of a donation package: "The foundation for the school has been poured, the bricks have been baked—now we need your help once more!" Everyone likes being able to see their impact.

Beyond accountability and fundraising, SuperSight can help humanitarian aid organizations maximize their effectiveness. The World Bank, for example, used satellite imagery from Orbital Insight to measure the physical size and growth rates of slums surrounding cities to gauge poverty levels in very large populations. Gathering this data would have been unfeasible through traditional on-the-ground surveys. Orbital Insight's analytics have also helped NGOs map large-scale refugee movements so that they know where, when, and how to best deliver aid.

One London company, Forensic Architecture, uses immersive technology to map war crimes. Their business name refers to the use of architectural evidence within legal and political processes, in their case to help prove cases of human rights violations. The company is the brainchild of Eyal Weizman, a professor of spatial and visual cultures at the University of London, who assembled a team skilled in LIDAR imaging and visual search algorithms to stitch together social media photographs of murky humanitarian situations and use them as evidence in legal proceedings. Their work uses SuperSight to investigate chemical weapons, extrajudicial murders, and environmental abuses. They've worked with the *New York Times* to disprove Russian claims of planted chlorine bombs in an attack in Syria by creating 3D models from news media footage; showed that the Greek Coast Guard was responsible for the drowning of forty-three Turkish asylum seekers by using long-range thermal cameras, weather data, and footage taken by one of the survivors, who had a waterproof camera taped to her wrist; and used machine learning to help train algorithms that recognize images of unexploded tear gas canisters around the world.

Animals also benefit from SuperSight—as long as you discount a zebra's right to privacy. To track populations of wildlife such as zebras and giraffes over the sprawling savannas of Africa, a nonprofit called Wild Me created an algorithm that helps recognize individual animals based on their specific pattern of spots and stripes, which, like human fingerprints, are unique. First, engineers train a neural network based on photographs of each giraffe they want to track. Then the company processes photos taken by snap-happy safari folks all over the park, who are unknowingly contributing to citizen-science research. The geotagged roaming data helps conservationists track herds, study family groupings, and collect migration data.

Computer vision can help endangered species, too. Poaching is still a big business, fiendishly difficult to curtail, and it's getting worse. In 2007, only sixty-two rhinos were killed by poachers, but in 2014 that number ballooned by a factor of twenty, to 1,300, according to the World Wildlife Fund. In 2015, the UN estimated that one hundred elephants were still being slaughtered *every day* in Africa. One of the reasons it's so easy for poachers to hunt and kill their game is that it's hard to track where they are; there's a lot of open savanna out there, and authorities don't know where to look. Luckily, drones and satellites have better eyes and tireless computational brute force.

Using drones equipped with infrared cameras, conservation organization Air Shepherd scans the savanna at night from above. It's an airborne version of the seafaring whale conservationists, Sea Shepherd, but their job is to alert rangers to the unexpected presence of warm bodies—in other words, unwelcome humans. The WWF has also developed a similar initiative, as part of their Wildlife Crime Technology Project, that spots intruders with thermal cameras on stationary poles along park borders and atop rangers' trucks.

Much of the tech we've been discussing was, unsurprisingly, funded and developed for military situational awareness, and is currently used to help remotely operated drones efficiently take out strategic threats. (I'm not allowed to even share what Space Force is doing!) But as I hope these examples show, SuperSight's planetary-scale applications can also be noble, useful, and inspiring.

Refugee camps, historical ruins, rhinos—it all feels a little far from home, right? Well, the same technologies also will let us do a lot of good in our own backyard (and front yard, and even on our own roofs).

9.3 Extreme world makeover

neural networks for generative design

We should all be using solar panels. Period. The average cost for a sustainable energy system has fallen about 70% in the last decade, from $5.86/watt to $1.50/watt, so it's a financial no-brainer. For no money down, you can finance an

Project Sunroof calculates how much homeowners can save with solar given the location, size, slope, and sun exposure of their roofs.

installation and start saving a hundred dollars a month in the first month, and even more if you live in the sun-saturated South.

So why aren't we? It's complicated! Math, logistics, taxes, and aesthetics all play a role. Many homeowners fear it will make their houses shiny and reflective like the Tin Man from *The Wizard of Oz*. The process of figuring out the number of panels in what size you need requires learning to "talk solar" in unfamiliar units like kilowatt-hours. And change always comes with risk, whether actual or just perceived.

The pro-climate mission of Boston-based company Energy Sage is to get people to electrify their homes. That means solar panels on your roof, an electric car, a home battery system, automatic blinds, and a smart thermostat that precools or preheats as you drive home. And they've partnered with us at Continuum to get potential customers more comfortable with the idea by showing them what an electrified version of their home might look like. Using publicly available Google Home satellite images, we size solar panels, digitally overlay them on clients' roofs, and then show them what their pad would look like from both the street and their neighbor's fence. We then take those images and pair them with data from Project Sunroof, a Google project that helps you work out the solar savings potential of your roof. Once you've seen the beautiful pictures of your electrified home and realized how much you're going to save over the years—and you have the visual and financial data in hand—it's a simple decision to go forward and make that change.

Other home improvement projects will benefit from a similar SuperSight-envisioned approach. Let's consider landscaping: another complicated, potentially expensive project with its own disorienting language, risks, and desperate need for pre-project visualization.

I met landscape designer Julie Moir-Messervy at an MIT pitch competition and was immediately intrigued with her mission: to give homeowners the confidence and tools they need to change their barren yard into a collection of outdoor living spaces. Her company, HomeOutside, helps people see new possibilities for their backyards using AI and computer vision. Once they've visualized their yard in a compelling way, the company makes it easy for them to make that vision a reality by hiring the landscape installer, getting materials delivered, and even helping spread the payments out over time.

Landscaping isn't just good for property values; greenscapes filter airborne pollutants that trigger asthma, help people recuperate faster from illness, reduce summer temperatures, and even lower crime. Proper native landscaping powers a dynamic system that helps out the bees and birds, who in turn pollinate trees and reseed plants. Southwest shade trees can reduce the need for air-conditioning, and northeast hedges cut down on winter winds—and heating bills. More trees mean more carbon capture—a ton over the lifetime of each tree—as they literally suck the bad stuff we produce out of the air while reducing runoff and erosion.

AI algorithms like scene segmentation automatically score every yard from Google Street View and satellite data, then a generative network selects plants, trees, and furniture to design a landscape that increases the home's value, beauty, and sustainability.

HomeOutside uses algorithms with an ecological bias to redesign residential landscapes in 3D.

But "most people don't do anything in their yards because they don't know where to start," Julie told me. "They don't know which plants to select and how to arrange them, or don't know how to install a landscape design and care for it over time." I was so inspired to work on the problem that I accepted a position on her board and got to work.

HomeOutside is training a generative adversarial network (GAN) to automatically compose beautiful and sustainable landscape designs, based on the thousands of designs (think of these as recipes) the firm has developed for clients over the last twenty-plus years. The company uses Google Earth Engine and photogrammetry to start with a 3D view of any address (US only, currently). The GAN architecture then uses one network (the Generator) to make a new design, and another network (the Discriminator) to judge or score the work. These two networks continue their iterative game, generating then scoring, until the discriminator judges that the landscape has a good composition: shade trees, natural pollinators, grass for playing, hardscapes/decks and furniture for gathering places, plant diversity, and so forth.

Companies that sell plants, furniture, lighting, and hardscapes are obviously interested in this type of "imagination engine" technology, because it bridges the conceptual gap between the current state of someone's garden and what could be—thus motivating many more people to make the dream real. It's not just great for the homeowners and outdoor retailers, either—it's great for the environment, too. But what the company's environmentally focused investors find most captivating about this project is the opportunity to change the landscape of entire neighborhoods at scale. *What if we could create a new national park across millions of backyards that stitch together places for birds and bees?* Every acre of forest absorbs about 2.5 tons of carbon a year. What if we turned neighborhoods into significant carbon sequestration zones?

I helped Julie and her team develop HomeOutside's grand plan to proactively redesign seventy million front yards, then work with Home Depot, Lowe's, Wayfair, IKEA, and garden centers to email their customers a 3D redesign of their yard. Customers simply go outside their home, open their phone, and, through the app's use of spatial world anchors, walk through an immersive animated landscape superimposed on their current yard. A time-lapse view from sunrise to sunset shows why the edible garden is placed where it is. The winter visualization explains the choice of new fir trees between their yard and the neighbor's. Spring flowers bloom with a cacophony of color.

Will people be alarmed by the idea of an algorithm proactively redesigning their yard, with new shade trees and naturally pollinating shrubs? It's not as if your front yard is private now, thanks to Google Street View. And if you are selling your home, you might decide against hiring the landscapers and just choose to post images of HomeOutside's makeover version instead to maximize your curb appeal.

Once this visioning technology is commonplace, lots of different fields will start taking advantage of it. Home Depot, for example, recently invested in a startup called Hover, which, after digitizing your home in 3D, visualizes and prices new paint, siding, and roofing materials. SuperSight will soon show the actual paint crew up on their ladders, finishing the last few brush strokes, so you get that delightful experience of a job just finished. Volkswagen might put a new Passat in your driveway, complete with the kayaks and mountain bikes it knows you love on top. And the company trying to sell you home and car insurance? They'll project a disaster scenario: solar panels fallen off, the shade tree hit by lightning, and your new Passat pummeled in a hail storm. Better buy the insurance before you repaint.

How will we interact with these types of immersive designs? With our SuperSight glasses on, will we point and place trees, or paint flowers from a palette of choices, like a 3D version of Photoshop? Will we select each plant from a vast menu of options for infinite control and customization, or will we just tell the system what we like so it learns our preferences, then proposes a single solution we'll love? I believe in the happy medium: that we'll largely prefer to see several "expertly composed" options and choose from among them, much as we do today when working with an architect, interior designer, or wedding planner.

Experts are usually so good at what they do that it's often a mistake to over-specify particular details. For example, you shouldn't tell an architect that you want a window exactly here, or an interior designer that you want this particular chair in a specific color in this corner. Instead, you express your opinions at a higher level of abstraction ("I want the room to feel more connected to the environment") or through describing a required function ("We want a vegetable garden"), and let them do the detailed work.

The same expert-guided interaction model will dominate our relationships with SuperSight AIs. For landscaping, we might ask for a more formal French garden with rectilinear layouts and exotic colorful plants, or a curvaceous organic design that prioritizes privacy from our neighbors. We might indicate a preference for an open space for play, or for a filled-in scheme with more space for a productive garden. And as we express these higher-level interests, our 3D landscape design will dynamically recalculate to match our preferences. With SuperSight glasses on, we'll be able to test our hunches faster by seeing reconfigurations immediately and in context, superimposed on our real home.

The jury is still out on whether HomeOutside will be able to use this technology to convince millions of homeowners to invest significantly in a sustainable landscape. The testing is promising, though; customers are delighted to see their yards reimagined and restaged. In the next five years, HomeOutside plans to use Google Earth and street view imagery in a generative AI tool to automatically redesign tens of millions of landscapes, with sustainable plants, shade trees, natural pollinators, and bird-friendly berries. If it succeeds, it will mean a million homeowners will plant at least 3 million new shade trees, like oaks and beeches, that will each capture 48 pounds of carbon a year as they mature. That's 14 billion tons of carbon sequestered over those trees' lifespans.

As one of the HomeOutside advisors summed up, "You are building the equivalent of a new national park—the National Park of us! Visualization tools like HomeOutside can persuade home-owners to reshape the American landscape."

That's the ultimate potential power of SuperSight: to help people envision and imagine a future that benefits themselves and the planet.

9.4 Speculative fiction for social justice

imagining a better world *affective neuroscience*

One of my favorite children's books is *If the World Were a Village*. It uses the metaphor of a village with a hundred people to describe the world's diversity, explaining how many of those people would be Caucasian, Asian, Hindu, rich, poor, owners of a car or cow, and more. By scaling the world's demographics into something that a child can fathom, the book provides a powerful new perspective on humanity for kids and adults alike. Our assumptions of what the world really looks like outside of our neighborhoods are so far off, and the need for empathy so large.

Understanding cultural and socioeconomic differences is critical. Our myopia drives fear and racial injustice, as well as economic injustice and healthcare disparities, yet we have very few tools for comprehending the universe beyond our narrow point of view. We chronically struggle to fathom large problems, which becomes an obstacle to taking those problems on.

Perhaps this is the most important role that SuperSight might play in our lives: helping us see the true scale of important problems, and motivating action to take care of our future selves.

By creating new views of our world, SuperSight can be used as a tool for social justice. Through using augmented reality, we can communicate the scope of societal problems that most of us conveniently and easily forget exist.

For example, one day you might be scrolling on Instagram past images of nature, food, and posing friends, when an arresting image of a new skyscraper in Toronto appears. It feels familiar, but also impossibly strange. If you've been to this city, you know their CN Tower reaches over 1,815 feet high and serves as an emblem for the area. But this new super-development dwarfs that building—it is more than two and a half times its height. When was this built? How is it possibly supported? How many elevators do you need to service that many hundreds of stories? And how did you know nothing about it?

Although it looks photorealistic, the image is a fiction designed to make a point. It illustrates how big of a tower would be required to house all of Toronto's poverty-stricken and homeless population. The building, dubbed the #unignorable tower, is sized to accommodate 116,317 people.

A piece of speculative architecture (even if it's virtual) is so much more vivid and communicative than a number. As the center of a campaign by United Way Greater Toronto to spread awareness of the issue of poverty and raise funds to combat it, the Unignorable Tower makes an invisible population and their staggering suffering visible in a powerful and understandable way, by giving scale to a problem that is hard for our brains to imagine—as well as the size of the solution required to fix it.

An impossibly large virtual building dominates the CN Tower, highlighting the need to house almost 117,000 homeless people in Toronto.

The goal of the campaign is to render the region's poverty, much like the gigantic building, absolutely "unignorable." And based on the social media buzz this speculative fiction garnered, it seems to be doing some good. But what if the campaign went bigger? What if this building were projected into the air with AR, similar to the way the Tribute in Light beams six blocks south of the Freedom Tower every September 11, reminding New Yorkers of the Twin Towers that used to stand there? The ability to spatially project images like these into city skylines is one of SuperSight's great opportunities

SuperSight can be an environmental-impact time machine. It can find historical shots of your current view to highlight the effects of global warming.

to provoke, shock, and motivate action—especially for invisible causes, those we've become accustomed to or choose not to see.

SuperSight also has the ability to show futures that hopefully will never happen. The technology can vividly show the negative consequences of our inaction—and therefore motivate change *now*. This is perhaps the most important function of augmentation: to help vividly imagine distant futures so we take steps to enable the preferable path today.

No issue has needed this long-term visualization more than climate change, since its effects are hard to see in our own backyards. We watch documentaries about receding glaciers in Alaska, the ravages of sea-level rise on Pacific island communities, and the bushfires burning out of control in Australia, but we need to see the effects closer to home to inspire the extreme urgency we should be feeling.

SuperSight can give us new eyes to see the environmental damage our collective actions are causing. It can do this in two ways: by showing us the past, and by revealing the future.

First, it can show us the natural beauty that we have destroyed to create the car-dominant environment to which we've become accustomed. Going back in time through our smart glasses, we can see the natural wetland banks of the Charles River in Boston before it was steam-shoveled in to create Commonwealth Avenue and the affluent Back Bay, and the Emerald Necklace in Florida before it was paved over for condo developments and roads.

Al Gore really nailed this approach in a way that made us care on an emotional level. I was at his TED Talk in 2006 where he showed spectacular places that had suffered great change due to global warming. You never would have imagined what some parts of the world used to look like: magnificent glacial valleys with massive flowing ice sheets were contrasted side by side with the same locations more than fifty years later, now receded, tiny versions of their previous selves. It was heartbreaking and transformational. What made these image pairs so successful was that they visualized what we were losing. Seeing makes us believe.

That TED Talk was so effective at helping people see the consequences of our actions that it went on to become the documentary and book *An Inconvenient Truth*. Gore made me—and many others—realize that all of these marvels were gone, and more would disappear if we didn't change our relationship to carbon.

Now imagine adding SuperSight to the mix. Through using augmented reality to see what the world around us used to look like, we can better understand the consequences of past actions, and with this wisdom, avoid repeating them. Glacial melting, coral reef bleaching, erosion, the deforestation of the Amazon—the current-day results of all of these environmental phenomena should provide sobering evidence to motivate change. While wearing your SuperSight goggles on safari in Africa or in a walk through the Redwoods, you could see the ghosts of now-extinct white rhinoceroses galumphing around with the elephants, or observe the tips of where some of the oldest sequoias once stood.

The second way SuperSight can help us see the consequences of our actions is by zooming us forward in time to witness a realistically bleak version of the

A visualization of Miami in 2050 generated by combining satellite imagery with climate data. Might this future image shock people into making a change today?

future—the world our children will live in if we fail to act. It can show us not only what used to exist where a given house or parking lot is located, but what the same scene might look like in ten, fifty, or a hundred years' time. Being able to see the future in this way is especially effective when trying to comprehend sea-level rise. Given how materials age, the effects of climate change, and the shifting economic health of the region, what might a generation from now see here?

In *A Christmas Carol,* Charles Dickens deploys a classic foresight technique to motivate his curmudgeonly Ebenezer Scrooge: a dream that shows him a dark vision of his future should he fail to change his path. We can use the same technique with SuperSight to reveal a future world impacted by the cumulative effects of a warming atmosphere, extreme weather, and a collapsing ecosystem. If we can project visions of this future into millions of SuperSight glasses, maybe we'll gain the fortitude and resolve to stop it from actually happening.

SuperSight-enabled speculative design provides new perspectives on the challenges facing us today and tomorrow. With it, designers, companies, and policy makers have access to an incredible set of communicative tools, and a publishing platform to help dreamers propose improbable and provocative futures.

The speculative design duo Dunne & Raby spoke at MIT in 2019 about their design treatise, *Speculative Everything*. In it, they propose a conceptual *what if* approach to the future that rails against typical design goals, such as a product being sexy or consumable. Instead, they focus on provocations and design fiction. Their philosophy sends students and teachers in experimental directions to generate weird "artifacts from the future" that are full of questions.

The close of their book has a short essay called "New Realities" that captures my design ethos—one I encourage you to embrace:

> To be effective, work needs to contain contradictions and cognitive glitches. Rather than offering an easy way forward, it highlights dilemmas and trade-offs between imperfect alternatives. Not a solution, not a better way, just another way. Viewers can make up their own minds.
>
> This is where we believe speculative design can flourish—providing complicated pleasures, enriching our mental lives, and broadening our minds in ways that complement other media and disciplines. It's about meaning and culture,

> about adding to what life could be, challenging what it is, and providing alternatives that loosen the ties reality has on our ability to dream. Ultimately, it is a catalyst for social dreaming.

The dreaming we must do, as we consider the limitations of our short-sighted brains, is about seeing into the future. About using the technology we are developing to provide a vivid rendering of what will be, if we don't change, or what could be if we do. If we can use SuperSight to summon the collective imaginations and creativity of all people and fuse it with what algorithms can synthesize, we will see great things and awful things, potential and consequences. Like Dickens's Scrooge, these dreams and their ghostly guides will be revelatory and shock us to change.

All of this suggests a new job description for the world of SuperSight: speculative designer. Through storytelling, SuperSight can bring the future to the present and help us act on it now. What dystopian futures can we imagine and prevent from coming into existence? What utopian ones might we conceptualize, illustrate, and then persuade stakeholders to fund and create? Speculative designers can help us see the scope and increasing scale of America's homelessness crisis, or the looming water shortages driven by our insistence on consuming animals over plants. With new eyes, we might better understand the urgency of reversing these crises, and start working together to ensure the future we want is the one that will come to pass.

SuperSight won't only help us see farther or through, small or large, fast or slow, backward or forward—it will force us to examine our own priorities and values, and inspire action and transformation.

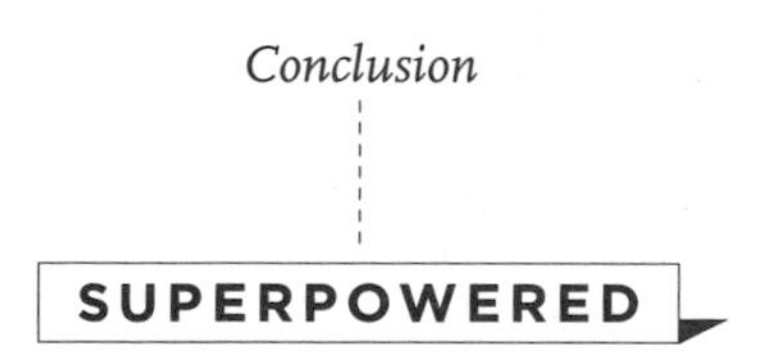

Conclusion: SUPERPOWERED

Humans have long dreamed of superpowers: X-ray vision to see through walls, the omniscience that comes from a bird's-eye view, or clairvoyance to see the future. Soon, these superpowers will come standard with your next pair of glasses.

The technology that makes this possible may not feel like technology at all, just like wearing sunglasses on a bright day—your eyes acclimate immediately and you only remember that you're wearing them when you take them off. But they will change the way we see the world, and in turn change the world we see. They'll provide access to an incredible amount of information and a connection to others that deepens our understanding and enriches our lives. But because this technology will be so embedded, the possibilities for a dystopia of surveillance, filter bubbles, and deepfakes also loom. I hope this book raises issues, sparks critical conversations, and conveys the sheer awe, wonder, and weirdness of the new world we're poised to create.

Just as mobile technology reshaped our lives in the last decade, SuperSight is poised to revolutionize human and computer interaction, turning it into something more invisible, ever present, and woven into the world. Human relationships, businesses, and entire industries will be transformed, and those business leaders who embrace this platform will position themselves and their organizations as prescient heroes. Superhuman vision won't just be for a select few like the Six Million Dollar Man, but instead will be a democratizing technology for all of us.

As we shape this new future, I'm committed to building SuperSight experiences that put *people* front and center—experiences that don't monopolize our attention, pollute social interaction, or remove us from the physicality of the world—or require an ad-driven business model. If you share these values, I hope you will pledge to do the same.

We've explored how society can and should respond to SuperSight, but let me suggest a few specific actions you can take now.

1. **Cast yourself as an anthropologist from another planet.**
   As the world changes around us, with new technologies and possibilities increasing exponentially, it's easy to feel overwhelmed. We must each decide if we are going to meet profound change with rigidity and skepticism, or with a learning mindset.

   To stay open and agile, I cast myself as a curious anthropologist when I encounter something new, interested in studying its effects on myself, coworkers, and family. I try to suspend initial judgment, appreciate the team and investment required to launch anything new, and experiment with a new product or tool as if it were new: considering what it might be good for, and any secondary effects its inventors didn't anticipate. Yes, this makes my family crazy, because we test (and often return) every voice-controlled, internet-connected, computer vision–based home appliance, toy, and robot. Each room of our home has ambient displays, touch-screen coffee tables, or data-projection countertops triggered by who's in the room, plus Post-its scrawled with the gesture or keywords needed to access these devices' features—but my family is learning about these new materials, too.

2. **Dive into gaming to see the potential for spatial computing.**
   If you're able, buy a VR headset and AR glasses to learn with. The best immersive experiences today are games like *Beat Saber*, *Half-Life*, *Skyrim*, and *Madden*, which will teach you about interaction conventions like ray casting to move, the power of spatialized audio, designing a good onboarding experience, and so much more. Dive in, and develop a point of view about what needs to be improved in these experiences to make them beneficial and compelling for different audiences in different contexts. If you were brought in as a SuperSight product advisor to a company, what prototype would you build first? What problem might SuperSight help solve? Which beachhead market would you land first?

Go deeper into content areas and applications areas you feel most passionate about—music education, architecture, or telepresence and collaboration experiences (I recommend a few on SuperSight.world). There are computer-vision apps that will assist with sports or hobbies, social skills like motivational interviewing, learning about street art, or ecological sustainability. Even if you aren't a coder, start sketching SuperSight experiences where it would be useful to know more or be motivated, or to collaborate, communicate, or persuade. (I mean literally start sketching your ideas, using tracing paper over photographs, or a similar onion-skinning app like Penultimate.) If there isn't an AR app for the situation you are passionate about, create one!

3. **Host a hackathon, participate in a design sprint, fund a prototype.** If you run a business or manage a product, you have many incentives for embracing SuperSight now. Would a 3D presentation of your product help customers or new employees better understand its manufacture and function? Where would you "place" this experience so more people encounter it serendipitously? (The answer is likely to be different for employee training versus customers.) Empower an internal research team, or engage an external design and innovation firm to prototype a few promising SuperSight services as an experiment. The best way to learn about this new world is through making and testing. Often the most rapid learning happens during week-long design sprints, or hackathons. I recommend partnering with a local university plus an AR software/hardware product firm and hosting a hackathon at your company. I maintain a list of AR events, project ideas, and design partners at SuperSight.world.

I began this book by marveling at the physiological wonders of the human eye, which, ironically, impressed even Darwin so strongly that he believed it could only have been crafted by intelligent design. As we've seen, these marvelous eyes of ours will soon receive an upgrade, evolving to possess mind-bending new capabilities. With augmented human vision, we'll all have faster access to knowledge, insights, and each other, as well as the ability to see the future—of ourselves and of the earth, for better or for worse.

Let's use these new, evolved eyes to imagine and see the world anew, and to paint a better future, one that feels plausible, compelling, and inevitable.

GLOSSARY OF KEY TERMS AND CONCEPTS

Some of these definitions drift into the land of conjecture or just simply my point of view, but that's why you bought this book, right?

Augmented Reality (AR) A layer of digital information superimposed over what you see in the real world. This is *not* virtual reality (VR), where your view is entirely filled with an opaque projection.

Rotoscoping The process of drawing over a filmed scene or character. In 1937, Walt Disney and his animators used the technique in *Snow White and the Seven Dwarfs*, projecting live-action movie images onto a glass panel to trace over and transfer each frame. Today, digital rotoscoping software uses motion tracking and onion skinning to create composite effects like glowing lightsabers. AR can now achieve these rotoscoped special effects in real time.

Digital holograms The output of SuperSight: projections that use the physics of diffraction to reproduce a 3D light field that creates a volumetric illusion. With AR, these holograms are dynamically painted into the real world with smart glasses or other translucent display technology.

Spatial computing A new way of thinking about how we organize digital information that gives priority to place. Information and services are anchored in the physical world or connected to tangible objects.

AR Cloud A giant content database linked to a 3D map of the physical world. It contains digital twins of objects and links to definitions, manuals, videos, historic layers, and more.

Metaverse Coined by author Neal Stephenson, "a shared, virtual space that's persistently online and active." Apparently this idea is why Facebook bought Oculus and is developing Horizons, probably why Google is developing Stadia, and why Epic Games developed *Fortnite*. In April 2021, Epic raised $1 billion to build the Metaverse.

Buena vista hypothesis The idea that better visual acuity in our vertebrate tetrapod ancestors from the Devonian period triggered the evolution of larger "planning" brains, as proposed by evolutionary biologist Malcolm MacIver. This begs the question: *How will the next evolution in augmented vision help humans level up?*

Field of view (FOV) The angle in front of the eye in an AR display that optically combines virtual layers with a view to the world. Larger FOV requires more computational power, resolution, size, weight, and energy (i.e., battery power). Humans have an FOV of 180 degrees. Today, the best AR glasses provide an FOV of 40 to 50 degrees.

Spatialized location and mapping (SLAM) A real-time computing algorithm that generates a depth map (also called a mesh or point cloud) of the world around you. AR glasses with this ability can position information in the environment, at the right scale, so it remains in that place even when you look away. Also called *spatial mapping*.

Occlusion An important feature of an AR system that allows objects in the world to appear in front of projections. It's required to have characters or information appear behind, for instance, a chair, or to be seen through a window in your home.

Spatial anchors An AR technique to position digital content in precise locations. For example, GPS coordinates or visual features used to position a sustainable landscape design in homeowners' yards, or topographical underwater maps over lakes and coastal waters for boaters. Distinct from *plane anchors*, which are used to position virtual content on any wall or floor plane.

Fiducials A visual mark designed to be computer readable, then applied to the world as a unique label (like a barcode or QR code). For example, a sticker or ball applied to a part of the human body to track its position in detail, as in motion capture.

Image anchor Any arbitrary image that acts as a fiducial to trigger an AR experience, like a QR code. It also provides scale and orientation. The thumbnail images in the sidebars of this book are examples of this. The SuperSight Instagram filter is trained on each of these images.

Receiver operator curve The key performance metric of a computer vision classifier that graphs the trade-off between sensitivity and precision. Sensitivity measures if the algorithm is able to identify something of interest. If it does, precision measures how accurate this prediction is.

Convolutional neural network (CNN) Algorithms were inspired by the connectivity patterns in neurons in the visual cortex; also known as deep learning networks. Once trained, CNNs are very efficient. Each layer, or convolution, is responsible for distinguishing features at different scales, from fine patterns to gross shapes.

Diminished reality The augmented equivalent of an ad blocker. Instead of adding more information to a scene, diminished reality strategically identifies and deletes content. The systems often use a neural network or an off-axis camera to "inpaint" or fill in what was behind the object or person that was removed.

Digital twin A digital copy of something physical, often with more data associated with it. For example, the twin of a plane engine might include not only the digital design model, but also performance stats, travel history, service records, training modules, and predictive algorithms for failures derived from other engines. SuperSight makes this information more legible and useful by positioning it on or around the physical object.

Feed-forward interfaces Auditory or visual cues that show you what a system plans to do, such as a backup beep on a truck. This predictive "what's likely to happen" information, akin to the probability cone of a hurricane's path, can be shown via SuperSight to increase safety.

Predictive feed-forward interfaces might be expressed as a risk cloud surrounding other drivers or intersections in the city.

Freehand gestural interfaces Sign language–like interaction modality used to communicate with our digital agents. We started interacting with computers using controllers like keyboards, then mice and pens. In the last decade, voice has improved and has dominated many daily interactions. Now computer vision systems can see fingers and hands to recognize gestures with incredible precision, even from a distance.

Gaze and commit An input model, akin to point-and-click with a mouse, that uses eye and head tracking to work out what a person is looking at to trigger an action. The commit can be done based on the duration of the stare (called dwell time), with a gesture like a snap, or via voiced input like a guttural. Using gaze suffers from *the Midas touch problem,* where everything you fixate on has the potential to trigger an unintentional action.

Multimodal input An approach to human–computer interaction that uses multiple input modalities for greater accuracy; how smart glasses will infer our intent. Touch, gesture, speech, affective signals, and even brainwaves each have their own shortcomings as inputs, but when fused together provide more effective and natural interactions.

Generative adversarial networks (GANs) An advanced AI approach that uses two neural networks to synthesize a painting, a better landscape design, a deepfake video, or a metahuman. One neural network (called the generator) makes something, then another (called a discriminator) judges and scores the work. If the scoring algorithm is reliable, the system improves over millions of iterations.

Pre-attentive processing A perceptual phenomenon where certain types of abstract information are interpreted by the brain in less than a second, in parallel and without any incremental cognitive load.

Pose detection A special type of neural network used to infer positions of a person's skeleton. We saw this technology in the AI coaching project in chapter two and the home health assistant Guardian Angel product in chapter seven. With a single camera we can now reliably estimate human pose and classify activity for health at home, for retail analytics, and for sports, even for many players who are partially occluded.

Design fiction A generative tool to imagine and explore possible future scenarios; also called *futurecasting*. Often these multiple parallel worlds are generated by extrapolating social and technology macrotrends or by combining the unknowns that are most salient to a company. Immersive technology like AR makes these possible worlds more tangible and vivid, and provides novel perspectives that reveal opportunities for innovation.

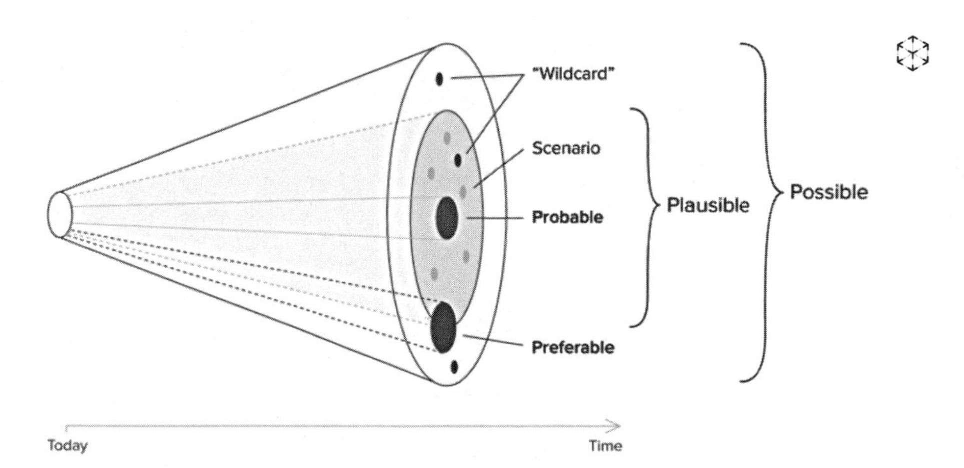

THE SIX HAZARDS OF SUPERSIGHT

Throughout the book, I identify six distinct *hazards of SuperSight*. Here is a grid of these concerns, and suggestions for how we might remedy each one.

HAZARD	HOW WE MIGHT ADDRESS IT
SOCIAL INSULATION Computer vision could trap you in your own personal view of the world, impeding your ability to connect with, understand, and empathize with others.	Allow people to synchronize the layers of information they see with each other: for example, people taking a college class might agree to see the same information as they walk through a national park or watch a musical performance. We might invent ad hoc gestures for two people to share what they are seeing with each other, like a "share your screen" for SuperSight glasses, or akin to a visual headphone splitter—perhaps executed with a fist bump gesture, or a reciprocal double-nod from a distance.
STATE OF SURVEILLANCE Pervasive cameras, both worn on our bodies and installed in everything from school hallways to doorbells, could provide governments and private companies unprecedented and incriminating data about individuals.	The solution here must be a legislative one. Civic leaders must establish new safeguards and laws to protect privacy and guarantee the right to be invisible and forgotten. We need new regulations that bolster visual privacy with onerous consequences, just like the General Data Protection Regulation did for data privacy in Europe. All SuperSight services must offer a purge button that gives anyone the right to see what visual data was collected on them and wipe it.

HAZARD	HOW WE MIGHT ADDRESS IT
⚠ COGNITIVE CRUTCHES Any assistive technology leads to the atrophy of skills we are no longer practicing: fire starting, handwriting, memorizing, map reading, driving, etc.	Install challenges and games in your smart glasses so your abilities—from just leading a conversation without prompts to making challenging clinical decisions or landing a plane in a storm—don't atrophy.
⚠ PERVASIVE PERSUASION In the age of computer vision, companies and brands will not only be able to see your search history and events calendar, they'll also be able to see what *you see*, and what catches your attention.	To protect us from what companies know and infer about us, profile information and preferences must be controlled by consumers, and only shared—potentially in exchange for personalization and convenience benefits—as the individual consumer sees fit. These preferences should be structured like an onion, with public preferences like "nut allergy" available to any restaurant, and more private layers like "credit score" only available as part of a transaction and not retained. Transparency and editability of this preference onion needs to be a new norm.
⚠ TRAINING BIAS Increasingly we are trusting our lives to autonomous systems, often without a clear sense of their accuracy or the data on which they were trained. Today many AI systems are opaque and fail to explain themselves and the precise reasons for their recommendations.	To ensure that people trust the judgment and actions of computer vision–driven medical, law enforcement, and driving systems, we must curate new, more inclusive datasets and stronger legal protections. There is an opportunity to create cross-company coalitions to aggregate and share this data to remove bias for everyone's sake. We must also include confidence scores and mechanisms for legibility and explainability with every recommendation.

HAZARD	HOW WE MIGHT ADDRESS IT
⚠ SUPERSIGHT FOR SOME Societal inequality becomes entrenched early, so we must ensure that we don't create a digital caste system where access to SuperSight is restricted to a select set.	We can do a lot to ensure access to SuperSight-related tools and knowledge for those with fewer resources. Governments and philanthropists should establish infrastructure that provides free access to computer vision and connectivity, like in a public library model, so SuperSight's educational powers benefit those who need them most. One of the most defining characteristics of technology is that it becomes more affordable. Smart glasses, too, will rapidly slide down this inevitable slope of democratization, from magical and inexplicable to commonplace and taken for granted.

DESIGN PRINCIPLES FOR SPATIAL COMPUTING

It's taken some time and debate with collaborators across various projects, but I've developed a set of general design principles for augmented spatial-computing experiences. I hope that these are useful to you, too, as you create experiences for learning, collaboration, telepresence, and other applications.

1. **Make sure you are solving a spatial problem.**
   If you don't need to ground information in the world, maybe you don't need AR. Before starting, think carefully about the problem you are trying to solve, and the information or experience you hope to deliver. Might another medium suffice? AR takes a lot to develop, and requires visible cues to communicate to people that it's available. Make sure that you are using the unique power of this new medium appropriately: you must anchor information in the physical world, onto an object or body, and that information needs to be better understood when rendered in 3D.

2. **Augment something tangible.**
   For example, Curiscope designed a T-shirt to teach kids about anatomy, where AR projects a beating heart seemingly inside your chest. By using a piece of clothing as the anchor, the positioning and effect are much more compelling than if you'd designed a disembodied heart floating in space.

3. **Don't overwhelm the visual field.**
   Use the *15% rule*: reserve 85% of the field of view for seeing the world, then start populating the remaining 15% with augmented content. This rule also applies to time. Try to make your content so context specific that 85% of the time, the display is clear.

4. **Reveal the unseen.**
   Literal labels on objects can be useful, but SuperSight can show so much more. Show something surprising, useful, or insightful like X-ray vision, galaxy- or micron-size scaling, super-slow motion, or a city time-lapse through the ages. Or visualize complex aggregate data like risk clouds.

5. **Use multimodal feedback.**
   Humans conflate modalities as a rule. For example, the smell of perfume is cross-interpreted with the bottle design, and sound design rubs off on cinematography. So start with combinations of haptics, sound, and visual cues for feedback. Go all in with modalities, then pare it back later if it's too much.

6. **Encourage natural gestures.**
   Certain gestures like *shhh*, *speak up*, and *look over here* are ingrained in our psyches. Instead of teaching people totally new ones, leverage the gestures with which we are familiar. For example: Google Earth VR uses the gesture of drawing your non-dominant hand to your face, as if you are peering into a crystal ball, to transition to street view.

7. **Direct attention to plot points.**
   The problem with being able to project critical information anywhere is that your audience can miss details while looking in the wrong direction. The solution, in SuperSight design as in narrative writing, is to draw increasing attention to salient details until your audience is compelled to turn and observe. Make sparks or use sound to draw attention.

8. **Use projected light for shared experiences.**
   If you are designing for a group experience, it may be easier to apply augmentation to the object itself with projected light versus making everyone wear smart glasses with a synchronized view. This ensures that everyone sees the same mixed-reality experience, and is confident that they are all seeing the same thing. For the MIT CityScope, we used projected heat maps over a LEGO city model to show walkability and

traffic congestion so a group of people could debate urban-planning options.

9. **Rotate objects, or subjects, to show off your dimensions.**
   Mercedes-Benz augments model cars as they turn on physical turntables to impressively show off new car details in AR. The *New York Times'* article "Augmented Reality: Four of the Best Olympians, as You've Never Seen Them" encourages you to walk around an AR figure skater suspended in the middle of a quadruple jump or a short-track speed skater racer leaning into a turn. As you walk around them, you appreciate the parallax and dimensionality of the AR scene because your view changes relative to the model.

10. **Lead with a strong POV.**
    Most 3D tools and AR interactions are burdened with too many options and overwhelming interfaces that control hundreds of parameters. Instead, use AI to make educated guesses to deliver value immediately. For example, start with a full room design, not a furniture catalog where placing each chair requires ten clicks. Generate a re-landscaped yard in full bloom, then refine the details after automatically showing something compelling; otherwise your customer may never get through the many small steps necessary to see the finale.

11. **Don't litter the world with your spatial junk.**
    When AR becomes ubiquitous, game companies, product makers, ecommerce brands, and individuals will be filling the built environment with virtual characters, instruction manuals, price tags, and spatial links to audio clips. As an AR designer, consider the scope of audience and persistence of the digital layers you're creating. Maybe they are reserved solely for a user's friends or a project team, and maybe they should fade at the end of the day or project.

12. **Design for episodic interaction.**
    Some of the best interactions in AR last only a few minutes, like the

virtual glasses try-on from Warby Parker. In less time than it takes to listen to a pop song, you gain confidence about what looks good on your face and can order a home try-on. You don't need to keep someone inside your app for hours—instead, encourage them to come back for short sprints again and again.

13. **Use diminished reality to restore focus.**
Don't forget to design for the things you can *remove* from view, not just add. SuperSight ad blocker, anyone? That should totally subvert Google's business model.

14. **Use AR to keep things current.**
Static mediums like print can

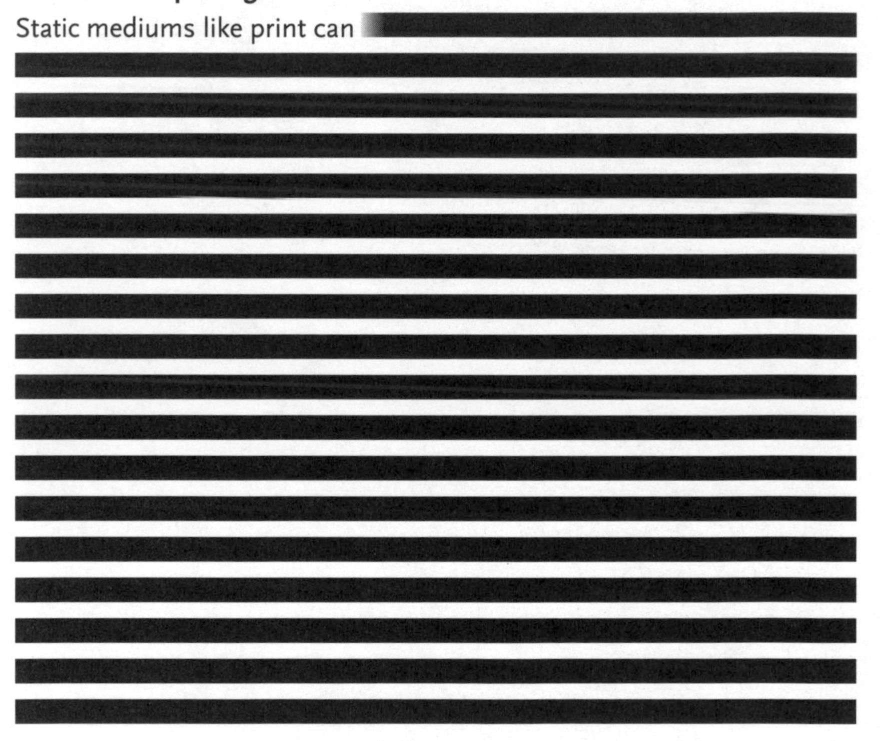

If you are actively designing AR experiences, I'm eager to learn what new principles you might propose. I'll keep a list on SuperSight.world/DesignPrinciples. It's a new medium; let's learn it together.

ACKNOWLEDGMENTS

All futurists owe a special debt of gratitude to the science fiction writers who anticipated and imagined the metaverse through books (Neil Stephenson, Philip K. Dick, William Gibson, Ernest Cline, Daniel Suarez) and the artists who animated these stories in film. Now, we are all truly living in sci-fi times.

Since I first crossed through the IM Pei arch outside the old MIT Media Lab in 1990, I have had the opportunity to swim in a world of ideas about a computation-rich future that pervades education, sports, creativity, transportation, food, cities, and opera. Thank you, Nicholas Negroponte, Muriel Cooper, Patti Maes, Hiroshi Ishii, Kent Larsen, and Roz Picard, for creating this place and community that has been such a wellspring of ideas and collaborators for nearly all of my adult life.

For this book specifically, I have enjoyed and learned so much from discussions with many idea partners and project collaborators: Neil Mayle, Joshua Wachman, Adrian Westaway, Pritesh Gandhi, Mark Schindler, Hani Asfour, Gerd Schmeita, George White, Simon Herzog, Josh Honaker, Rick Borovoy, Matt Cottam, Gilad Rosenschweig, and Hari Nair.

Thank you, Jordan Goldstein and David Gensler, for the Gensler fellowship year to dive deep into the future of responsive and sentient workspaces. To Ari Adler, Todd Vanderlin, and the talented team at IDEO, I appreciate your ongoing friendship and collaboration on the gestural interfaces that will be so important in the coming years.

To my most recent collaborators at EPAM Continuum, you have been especially delightful to work with. You are full of heart and humility, soaked in science, and appreciate the complexity of human-centered design and systems thinking. Thank you, Gaurav Rohatgi, Buck Sleeper, Chris Michaud, Jonathan Campbell, Kristen Heist, Eric Bogner, and so many others, for proposing a title of futurist and for encouraging my AI and computer vision experiments in sports coaching, home healthcare, and immersive computing.

Chris McRobbie, thank you for this book's cover design and chapter AR animations, for crystallizing SuperSight design principles, and for being a steadfast design partner as we explore and push on the limits of this new "material" of mixed reality.

For the marathon of book-making, I'm grateful to Todd Shuster, my stalwart agent for his partnership, and for the introduction to Georgia Frances-King, my exuberant structural editor and collaborator. Leah Wilson, you have been an amazing editor, distilling and sharpening my long-winded stories and overly optimistic foresight. Thank you for your dedication to the project and for coordinating the talented team at BenBella who have been so attentive and delightful to work with.

Researching this book, I learned from many subject matter experts who helped me consider how AR might impact their domains, including Amy Traverso about cooking, Jinha Lee about 3D spatial collaboration tools, and Susan Conover about computer vision for dermatology. Thanks especially to Julie Moir Messervy, and the passionate team at HomeOutside, for taking me to school on landscape architecture and the huge opportunity to use GANs and AR to help people imagine a more sustainable world.

I'm perhaps most proud of, and thankful to, my students at MIT and the Copenhagen Interaction Design School, who boldly build prototypes to "think with," have over the years matured into extraordinary inventors, designers, and entrepreneurs, and are now contributing to teams developing some of the most innovative technology experiences at Google, Microsoft, Apple, PTC, and more.

Thank you, family, for tolerating my travel for teaching and talks, disappearing to New Hampshire for writing benders, and testing all my prototypes—especially Adam, who may never be the same after all the AR experiments.

INDEX

Bold indicates tags.

T

ART CREDITS

Chapter title page and part opener page illustrations © Chris Robbie
Pages 43, 141 (bottom), 158, 170, 191, 192, 223 © Chris Robbie
Pages 5, 261: From the United States Patent and Trademark Office
Pages 29, 36, 127: © Micah Epstein
Pages 36, 168: Courtesy Tellart
Page 46: From "IO Bulb and Luminous Room," by John Undercoffler et al, MIT Media Lab
Pages 58, 67, 118, 150 (top), 152, 182: Courtesy EPAM Continuum
Page 70: © Ehsan Hoque
Page 150 (bottom): Claes Oldenburg and Coosje van Bruggen
Spoonbridge and Cherry
1985-1988
aluminum, stainless steel, paint
354 x 618 x 162" overall
Collection Walker Art Center, Minneapolis
Gift of Frederick R. Weisman in honor of his parents, William and Mary Weisman, 1988
Art © Claes Oldenburg and Coosje van Bruggen
Page 172: Courtesy NTT Human Interface Laboratories
Page 189: Andrea Day on Unsplash
Page 198: Courtesy Dina Katabi, photo ©Jason Dorfman
Pages 200, 201: Courtesy EPAM Continuum; image by Marnie Chang
Page 246: Courtesy Home Outside; original image from Google Street View
Page 247: Courtesy Home Outside
Page 291: Author photo by Lavin Agency
All other images used are fair use or from the public domain.

ABOUT THE AUTHOR

David Rose, MIT lecturer, inventor, and five-time entrepreneur, draws on culture, design, travel, and music to envision future products and businesses sparked by the next generation of technology. His last book, *Enchanted Objects,* is the definitive book on designing the Internet of Things. David wrote the seminal patent on photo sharing, founded an AI company focused on computer vision, and was VP of Vision Technology at Warby Parker. He is known for translating complex technologies into delightfully intuitive new products and consulting with businesses on how to thrive during digital disruption.

David's work has been featured at the New York Museum of Modern Art, covered in the *New York Times, WIRED,* and the *Economist,* and parodied on *The Colbert Report*. His home was featured in a *New York Times* video "The Internet of Things" about inventions that incorporate magic into everyday objects: a Google Earth coffee table that responds to gesture, Skype cabinetry in the living room, and a doorbell reminiscent of Mrs. Weasley's clock that rings when a family member is on their way home. He even got John Stewart to belly laugh when he was a guest on *The Daily Show*!

dlr@mit.edu
SuperSight.world